Antoine Lavoisier
Oxygen, Acids, and Water

Antoine Lavoisier

Oxygen, Acids, and Water

Eight chapters from the
Elementary Treatise on Chemistry

Translated by Chester Burke and Matthew Holtzman

Edited and annotated by Howard J. Fisher

Green Cat Books

An imprint of Green Lion Press
Santa Fe, New Mexico

Manufactured in the United States of America.
Published by Green Lion Press
Santa Fe, New Mexico
www.greenlion.com

Green Lion books are printed on acid-free paper.

Set in 12-point Arno Pro

Printed and bound by Sheridan Books, Inc., Chelsea, Michigan.

Cover: Detail from Jacques-Louis David, Portrait of Monsieur Lavoisier and His Wife (1788).

Photo credits: p. 20—CNAM, Conservatoire National des Arts et Métiers, Paris/Bridgeman Images; p. 33—© Musée des arts et métiers–CNAM, Paris/photo J-C Wetzel.

Cataloguing-In-Publication data:

Lavoisier, Antoine
Oxygen, Acids, and Water: Eight Chapters from the Elementary Treatise on Chemistry / translated from the 1793 edition by Chester Burke and Matthew Holtzman
Edited and annotated by Howard J. Fisher

Includes abridged text of Antoine Lavoisier's Elementary Treatise on Chemistry, index, introductions, biographical sketch, glossary, and notes.

1. Lavoisier, Elementary Treatise on Chemistry. 2. History of Science.
3. Chemistry.
I. Lavoisier, Antoine-Laurent de (1743–1794) II. Burke, Chester (1953 –)
III. Holtzman, Matthew (1978 –) IV. Fisher, Howard J. (1942–)

ISBN-13: 978-1-888009-51-4 (softcover binding)

Library of Congress Control Number: 2019933371

Table of Contents

The Green Lion's Preface
Note on the Translation

The present little volume is a partial realization of a long-standing goal of the Green Lion: Lavoisier's magnificent *Treatise* has too long been hobbled by an archaic English translation, and we have hoped to set it free. Our aim, however, goes far beyond modernizing the text. Because the present translation is different from other renderings of scientific works that readers may have encountered, we think it appropriate to say what we think a Green Lion translation should accomplish.

There is a lot of variety in the Green Lion translations, but one feature they hold in common is a deep respect for the original text. We have all seen plenty of translations that are tinted, and sometimes tainted, by an evident feeling on the part of the translator that he or she knows the author's intent well enough to rephrase or even paraphrase passages that appear awkwardly or obscurely worded. These efforts certainly have their place, but that's not where we want to go. We are especially interested in the odd or clumsy expression, because those oddities often bespeak a remarkable and original idea for which the author does not yet have a clear expression. We want readers to be able to watch a profound thinker struggle to think through a new and difficult thought.

We are committed to a translation of this text that reveals and traces Lavoisier's thought and voice as much as possible. We want to be sure not to elide over nuances of diction, imagery, and metaphor that may bear on precisely the way Lavoisier was thinking about the phenomenon he is describing. We pay attention to maintaining Lavoisier's voice, avoiding phrasing that verges on paraphrase, and keeping a little closer to the structure of the French where we do not see a clear advantage in departing from it. We value a translation that is scrupulous about bringing through the original author's written expression, including images and subtle resonances, even at the expense of more awkward and unfamiliar wording, leaving room for generations of students, readers, and scholars to dig out interesting angles. The translation should stand on the solid ground of the text, perhaps most of all when we are not sure why that ground has the shape that it displays. We write for those who study the poetry of Lavoisier, as well as those who study his science.

In writing and refining the text and notes for this Green Cat Module, the Green Lion editors and translators have been immeasurably aided by our long involvement, as students and teachers, in St. John's College's unique approach to the study of scientific texts. For over fifty years, on two campuses, every freshman has spent several weeks of the spring semester in a close study of Lavoisier's *Treatise*—a study that involves not only reading and discussion but also working through Lavoisier's chief experiments in the laboratory. The editors and translators have thus benefited from St. John's deep and rich body of experience that we can justly claim to be unequalled anywhere in the world.

For this translation the Green Lion contracted with Chester Burke and Matthew Holtzman, two faculty members at St. John's who, in addition to an impressive command of the French of Lavoisier's time, have many decades of experience reading Lavoisier's *Treatise* with nonspecialist undergraduate students. The Green Lion editors, Howard Fisher, Bill Donahue, and Dana Densmore, have all also had extensive experience reading and helping students read Lavoisier—all three of us starting in the 1960s. Fisher and Burke were important early and ongoing guides to the laboratory program on the Annapolis campus, and Donahue was Director of Laboratories for many years on the Santa Fe campus.

It has been our joint effort to bring through Lavoisier's thought in all its lucidity, preserving the excitement of this moment, a moment on the cusp of making new sense and order out of a jungle of chemical phenomena. The resulting translation is faithful to the text in the ways described above, while producing a version that reads smoothly in English. We are honored that Burke and Holtzman have entrusted us with their work, and we are delighted to present it to a larger public, enhanced by Fisher's illuminating notes.

Dana Densmore
William H. Donahue
for the Green Lion

About Antoine Lavoisier
26 August 1743–8 May 1794

Lavoisier's accomplishments in chemistry were numerous and consequential; but central to them all was his appreciation of the importance of *weight* in chemical reactions—an insight which changed chemistry from a qualitative science to a quantitative one.

A quantitative science depends on measurement; and in the eighteenth century, instruments capable of precision performance were enormously expensive to build. Because Lavoisier had inherited a large fortune, he possessed the means to construct and maintain a large and sophisticated laboratory and furnish it with superbly-crafted scientific apparatus.

Lavoisier married Marie-Anne Pierrette Paulze in 1771. She was to become deeply involved in Lavoisier's scientific activities, assisting in the laboratory, documenting experiments and apparatus in sketches and notes, and translating the scientific writings of English researchers. It was she who prepared the engravings for the 1789 publication of the *Elementary Treatise on Chemistry*—engravings that are reproduced in this Green Cat Module.

Lavoisier's principal achievements and fame lay in the field of chemistry. Nevertheless, throughout his life he devoted substantial efforts towards social good, promoting improvements in agriculture, air and water quality, and public health.

In 1768 Lavoisier purchased a share in a company that collected customs, excise, and other taxes under royal charter. Although the income from this source helped finance his scientific work, Lavoisier's association with tax collection later made him a target of the revolutionary powers; he was put to death on 8 May 1794 at age 50.

The mathematician Lagrange memorialized Lavoisier's execution with these words: "It took them but an instant to cut off this head, and one hundred years might not suffice to produce another like it." A year and a half after his death, Lavoisier was exonerated by the French government.

Editor's Introduction

Lavoisier's *Elementary Treatise on Chemistry* knits together an astonishingly wide range of topics, including the fundamentals of his *caloric* theory, an encyclopedic survey of known chemicals, and a meticulous description of chemical apparatus and techniques. It is a sprawling work in many respects; but the first eight chapters stand remarkably on their own. Together they constitute a train of argument leading from the apparent nature of acids to Lavoisier's demonstration that water is no simple substance, but a composition of oxygen (the "acid-maker") and hydrogen (the "water-maker"). The title "Oxygen, Acids, and Water," though not Lavoisier's own, conveys, I think, the focus of his thinking in these chapters.

Not so very long ago, a typical college introductory course in chemistry might be experienced as a dismal mass of facts, imperfectly organized according to principles which themselves rested on imperfect foundations. By contrast, what a splendid edifice was offered by *physics*, which from the outset presented a complete logical structure based on simple, universal, and intelligible laws! The example of physics did much to cement the extraordinarily influential idea that knowledge becomes *science* when, and only when, it is organized in the form of laws—preferably, mathematical ones.

Academic chemistry during much of the twentieth century did not stand up very well to such a criterion; and "physics envy" was an affliction with which theoretical chemists, not to mention social scientists, had often to contend. But chemistry springs from a tradition that cultivated a very different conception of knowledge: knowledge as *direct apprehension* more than subordination to demonstrative logic.

Reading Lavoisier's *Elementary Treatise on Chemistry*, we feel his abiding awareness that *matter has powers*; and we can participate in his desire to know those powers firsthand. Laws—especially laws aiming for universality—are largely beside the point; for Lavoisier's awareness is far more attuned to the distinctiveness of different kinds and classes of substances.

For that reason it is rewarding to pay attention to Lavoisier's verbs. When does he use terms like "release" (*dégager*), as distinguished from "remove" (*enlevèr, retirer*) or "extract" (*extraire*)? Lavoisier seems constantly trying to infer exactly what each substance is doing in relation to the other substances in each chemical reaction. When he cannot deduce, he *conjectures*—because his interest is in figuring out how each substance acts, more than in demonstrating how it conforms to an overall pattern. To be sure, he loves those large patterns when he finds them; but they are always welcomed as gifts, as prized fruits of study and labor—never as the substance of knowledge itself.

Studying Lavoisier, like studying the early works of any science, will be most rewarding if we can temporarily set aside concepts that our schooling offered to us ready-made, and let them instead become live questions. By doing so we can appreciate, and even take part in, the working-out of those concepts by the great founding thinkers.

Consider, for example, Lavoisier's understanding of heat as being a *substance*, which he calls "caloric"—a view which modern textbooks too often treat with condescension or even ridicule. Heat, as we all now know, is no substance, but rather a form of energy. But readers who can step back from "what we all now know" are regularly astonished to discover just how strong was Lavoisier's evidence for the existence of caloric, and what a rich explanatory power that concept had in Lavoisier's hands. It does no honor to later investigators if we think the theories they superseded were merely silly. If we really want to appreciate the modern concept of heat, we will want to understand the potency and utility of the theory it supplanted.

Along with ready-made concepts, we will do well to forego the standardized *terminology* in which those concepts are embedded. This is particularly appropriate for Lavoisier, as *names* were of central importance to him. The "Preliminary Discourse" which begins this volume is his compelling call to name substances according to *what we actually know* about them, not just what convention dictates or what practice has ordained.

As knowledge becomes more and more widely accepted, technical terms tend to migrate into ordinary speech, even to the point where we may forget that they *are* technical terms—and that they therefore embody a host of assumptions and suppositions that were once open questions. We, to a far greater extent than Lavoisier's contemporary readers, need to be aware of this.

Consider the term "molecule," for example, which for us denotes a definite cluster of fundamental substantial units. But Lavoisier's *molécules* are, literally, no more than *small masses* of the substance in question; he has no reason to suppose that these bits of matter are equal to one another, even in the same substance. Such intimate details of the constitution of bodies appear to be out of reach of Lavoisier's methods; but readers are invariably surprised and delighted to discover how much one *can* know about materials, and how interesting they can be, even when their minute constitution remains a mystery.

Commensurate with Lavoisier's striking scrutiny of the vocabulary of chemistry was the scrupulous attention he devoted to its physical apparatus. Chemistry, like other sciences, cultivates a highly disciplined brand of experience that requires highly specialized equipment to produce. Lavoisier worked at a time when not only the basic concepts of chemistry were in flux, but even its instruments and procedures were not standardized. While certain articles of glassware—retorts, flasks, and the like—had long been associated with chemical and alchemical work, even these had to be made more or less on an as-needed basis rather than produced as stock items. The balance had long been used in trade and commerce, but Lavoisier made it an indispensable instrument for chemistry by demonstrating the need for greater precision than ordinary instruments could deliver. The exceptionally capable Fortin balance (page 47), which he commissioned, brought that familiar appliance to new standards. On the other hand, instruments like Lavoisier's gasometer (page 88) were wholly new.

To what degree of knowledge can chemistry realistically aspire? In contrast to the demonstrative paradigm, Lavoisier knows that conclusive answers are not the only valuable ones. His science is not a "method" that guarantees truth; instead, as we shall see, it generates *experience* that can either clarify or overthrow predominating conceptions.

In this Module I have made use of a few formal devices that may need explanation. I have supplied paragraph numbering, in part to facilitate cross-references. Thus, for example, the number 3.10 in parentheses— (3.10)—directs the reader to consult paragraph 10 in Chapter Three for supporting remarks or other material related to the topic under immediate discussion. A few of Lavoisier's paragraphs are omitted in these selections; these are indicated by a trio of asterisks (* * *), and the paragraph numbers jump accordingly.

I have endeavored scrupulously to distinguish between Lavoisier's narrative and my commentary. Within each Chapter, his text always occupies the top of the page. My comments appear in smaller type at the bottom of the page, beneath a separator line, and keyed to each numbered paragraph—in most cases also to a particular phrase. Lavoisier's own footnotes appear beneath his text, but above the separator line.

Illustrations in the text are the ones created for the first (1787) edition of the *Treatise* by Marie-Anne Pierrette Paulze—Mme. Lavoisier. These were generously provided by Dr. JoAnn Palmeri, Curator of the History of Science Collections at the University of Oklahoma, and carefully prepared for publication by William Donahue. Illustrations in the comments are from various sources.

The editor wishes to thank Caroline Mason of the Los Alamos National Laboratory for sharing a modern chemist's insight into the combustion of phosphorus, described in Chapter Five. Chester Burke and Matthew Holtzman provided the new translation of Lavoisier's text—the first English rendition since 1790—for this volume. They are now completing the translation of the entire *Elementary Treatise on Chemistry*.

Howard J. Fisher
January 2019

Antoine Lavoisier
ELEMENTARY TREATISE ON CHEMISTRY
Part I

Preliminary Discourse

0.1   When I first undertook this volume, my only aim was to further develop the paper I read at the public meeting of the Academy of Sciences in April, 1787 on the necessity of reforming and perfecting the nomenclature for chemistry.

0.2   But while I was engaged in this work, I came to appreciate, better than I ever had before, evidence for the principles that the Abbé de Condillac set out in his *Logic* and in other works. Condillac has established that "we only think with the assistance of words," that "languages are true analytical methods," that "algebra, being the simplest, the most precise, and the most suitable to its object among all modes of expression, is both a language and an analytical method"; and, finally, that "the art of reasoning is, at bottom, nothing else but a well-constructed language." And so it turned out that while I thought I was only focused on nomenclature, and while my sole intention was to perfect the language of chemistry, my work, in a manner I could not resist, imperceptibly transformed in my hands into an elementary treatise on chemistry.

0.3   The impossibility of isolating nomenclature from science and science from nomenclature arises because every physical science is necessarily composed of three things: the series of facts which constitutes the science; the ideas that recall those facts; and the words that express those ideas. Now the word should give rise to the idea; the idea should depict the fact. Word, idea and fact are three imprints of the same seal; and because words preserve

---

0.1 *the paper I read ... in April, 1787*: Lavoisier refers to the first section (pp. 1–25) of *Méthode de nomenclature chimique* (Paris, 1787), which he authored jointly with de Morveau, Berthollet, and de Fourcroy.

0.2 *the Abbé de Condillac set out in his Logic*: Condillac, *La logique* (Paris, 1780).

and transmit ideas, one cannot perfect the language without perfecting the science, nor the science without the language. And no matter how certain the facts, no matter how sound the ideas that the facts produce, they would convey nothing but false impressions, were we not in possession of exact expressions to render them.

0.4   The first part of this treatise will provide those who carefully consider it with many demonstrations of these truths, but since I have been obliged to follow a plan for the treatise which differs essentially from the one currently adopted in all works of chemistry, I should explain the reasons that led me to do so.

0.5   It is a well-established principle, a principle recognized as quite generally applicable in mathematics and, indeed, in all branches of knowledge, that instruction can only proceed from the known to the unknown. In early childhood, ideas arise from needs. The perception of our needs produces an idea of objects that could satisfy them, and imperceptibly one idea after the next arises from our sensations, observations, and analyses, and all of these ideas are tied to one other in such a way that an attentive observer could, up to a certain point, even recover the thread that connects them. This chain of thoughts constitutes the sum of what we know.

0.6   We are much like children with respect to any science we have taken up for the first time, and we must follow precisely the path that nature follows in forming a child's ideas. For a child, ideas are the effects of sensations, sensations produce ideas, and the same should be true for all those who devote themselves to the study of the physical sciences: their ideas ought to be the consequence, the immediate result, of an experiment or observation.

0.7   I would like to add that whoever embarks on a career in the sciences is in an even less advantageous situation than the child acquiring its first ideas. If the child is mistaken about the helpful or harmful effects of objects in its environment, nature provides it with ample opportunity for correction. At each moment, experience amends a child's judgment. Privation and pain follow false judgments; joy and pleasure follow sound ones. It does not take long to become a very respectable thinker with teachers like these, for one quickly learns to reason well when the penalty for reasoning badly is pain or hardship.

0.6 *ideas ought to be the consequence ... of an experiment or observation*: Knowledge in the sciences should begin with *actual experience*—not with, say, doctrines inherited from tradition, or (as he will discuss in the next paragraph) with ideas we just come up with on our own.

0.8 But this is not at all true in the case of the study or the practice of science. The false judgments we make have no bearing on our existence or well-being, and no physical concerns oblige us to revise them. On the contrary, imagination (which constantly tends to carry us beyond the truth), as well as self-love and self-confidence (which all too often motivate us), tempt us to draw conclusions which by no means follow immediately from the facts, so in some sense we have reason to want to deceive ourselves.

0.9 It is hardly astonishing, then, that throughout the physical sciences suppositions so often take the place of conclusions, that these suppositions, handed down from one age to the next, should grow increasingly imposing as they acquire the weight of authority, or that they should, in the end, be adopted and regarded as fundamental truths by even the greatest minds.

0.10 There is only one method for preventing errors like these. We must eliminate, or at least simplify, our inferences whenever possible, since we are alone responsible for the inferences we make, and so they are the only things which can mislead us. We must constantly refer our inferences back to the touchstone of experience. We must only retain a record of the facts, since facts are the only things that nature gives us and so they cannot deceive us. And we must search for truth only among the natural connections between experiments and observations, like mathematicians when they are working towards the solution to a problem. For mathematicians merely arrange the givens and reduce all lines of thought to such simple steps and such straightforward judgments, that they never lose sight of the evidence that guides them.

0.11 Convinced by these truths, I adopted the following rules: to proceed only from what is known to what is unknown, to draw only such consequences as fall out immediately from experiments and observation, and to arrange the truths of chemistry in the order most appropriate for facilitating the understanding of beginners.

0.12 It would have been impossible for me to pursue this plan without departing from standard approaches. For almost every chemistry course and treatise

---

0.8 *false judgments [in science] have no bearing on our existence or well-being*: False ideas about everyday matters are swiftly corrected by casual experience. But the experience that science cultivates is decidedly restricted and often highly contrived. An erroneous opinion concerning the result of touching a flame will be rectified at once; but a false opinion about whether water is a simple or a compound substance does not affect everyday life and is unlikely to be corrected without the construction and use of elaborate specialized apparatus.

makes the mistake of presupposing at the outset knowledge that students or readers should only acquire much later in their studies. They almost all begin by treating the elements of bodies and explaining the table of affinities without realizing that this requires them, from the very first day, to present the fundamental phenomena of chemistry, to use expressions which have not yet been defined, and to assume that the student or reader has already acquired the very knowledge that the treatise or course is meant to impart. And because of this we accept it as a fact that little can be learned in a first course in chemistry; that a year of study is hardly enough to develop an ear for the language of chemistry or an eye for its instruments; and that it is almost impossible to become a chemist without three or four years of preparation.

0.13   But these difficulties pertain not so much to the nature of the subject as to the manner in which it is taught; and once I understood this, I decided to set chemistry on a course that seemed to me more in accordance with nature. I am aware that in order to avoid one kind of difficulty I have fallen into others, and that it would be impossible to avoid them all. But I believe that those

---

0.12 *the mistake of presupposing at the outset knowledge that students or readers should only acquire much later*: Aristotle distinguished between *the order of being*, which articulates what matters are primary or secondary in themselves, and *the order of knowing*, which identifies what matters are more and less immediately accessible to us. Lavoisier has been arguing, in effect, that learning must proceed according to the order of knowing; but traditional chemistry treatises have endeavored to present their material in a sequence that they consider to reflect the order of being. Not only does this constitute poor educational practice in Lavoisier's view, but it distorts the character of science, which—since it proceeds from the known to the unknown (0.5)—must necessarily reflect the order of knowing.

0.12 *the elements of bodies*: For example, the traditionally-accepted elements earth, air, water, and fire. To begin by discussing these is indeed an example of presupposing at the outset knowledge that can only be acquired later. For as he will note in (0.21), the determination that a particular substance is "simple" requires continual experimentation and, in a way, can never be regarded as altogether complete.

the table of affinities: The tendency of one substance to combine with another is called its *affinity* for that substance. It was customary for treatises on chemistry to present this information in the form of tables such as Geoffroy's of 1718, a portion of which is shown here.

TABLE DE MR GEOFFROY en 1718.

an ear for the language ... an eye for its instruments: Here Lavoisier seems to express the essential skills of the chemist; for the *instruments* are the means of gaining the experience on which chemical knowledge is based (0.6), while the *language* is practically synonymous with the organized body of knowledge itself (0.3).

which remain may be ascribed to imperfections in the science of chemistry rather than to the direction I have prescribed for it. It is difficult and confusing to find the connections between the facts of chemistry, because the series of chemical facts is riddled with gaps. Unlike elementary geometry, chemistry is not a complete science in which all parts are closely bound to one another. But at the same time, it is currently progressing so quickly, and facts are organized so felicitously under the modern doctrine, that we can hope, even in our own time, to see chemistry approach the highest degree of perfection it can reach.

0.14   These strict rules, from which I have never deviated, and which keep me from drawing any conclusions aside from those experiments show me, or from speaking when the facts are silent, do not permit me to consider in this work the branch of chemistry most likely, perhaps, to become an exact science someday: the branch devoted to chemical affinities or elective attractions.

* * *

0.16   It will be surprising not to find a chapter on the elementary and constitutive parts of bodies anywhere in this elementary treatise on chemistry, but I will say here that we are attracted to the view that all bodies are composed of three or four elements because of a prejudice inherited from Greek philosophers. The supposition that four elements, in various proportions, compose all known bodies is a pure hypothesis conceived long before anyone had the first notion of experimental natural philosophy and chemistry.

* * *

0.20   As I see it, all discussions of the number and the nature of the elements must be considered purely metaphysical because they involve the resolution of underdetermined problems, problems admitting of infinite solutions, none of which, in all likelihood, are in accordance with nature. So I will only

---

0.14 *elective attractions*: an alternative name for *affinities*, expressing a substance's tendency to combine with certain substances in preference to others; from "election" as *exercise of choice according to preference*.

0.16 *the elementary and constitutive parts of bodies*: A strong tradition, dating back to Aristotle, dictated that treatises on chemistry should identify a number of "elementary" substances, from which all other substances were supposed to be formed. Aristotle had singled out *earth*, *air*, *water*, and *fire* (representing, in turn, the four possible pairs of *moist* or *dry* and *warm* or *cold*). The 16th-century alchemist Paracelsus identified *sulphur*, *mercury*, and *salt* as elements.

say that if we mean, by "elements," the simple and indivisible particles that compose bodies, then it is likely that we are unacquainted with any elements. But if, on the other hand, we attach the terms "elements" or "principles" to the idea of the last result our analyses of bodies have obtained, all substances which we have not by any means been able to decompose will be elements for us.

0.21 We cannot be sure that the bodies we regard as simple are not themselves composed of two or even a great number of principles. However, if the parts of a body never separate, or, rather, if we can find no way to separate them, we should regard that body as simple and should not consider it composite until experiment and observation supply proof.

0.22 These reflections on the way ideas should develop apply naturally to the choice of words for expressing those ideas as well. Guided by the work that de Morveau, Berthollet, de Fourcroy and I produced together in 1787 on chemical nomenclature, I have designated simple substances with simple words whenever possible, and I was obliged to name them first. As you may recall, it was our aim to retain all common or customary names for substances. We only permitted ourselves to change a name in two cases: first, in the case of newly discovered substances which had either not yet been named, or had been named so recently that the names had not yet gained common currency; and in the case of terms, whether ancient or modern, which seemed to us to suggest obviously false ideas, as when, for instance, a term could cause confusion between its referent and another substance endowed with different or even contrary properties. In cases like these we felt no qualms about substituting other terms, usually derived from Greek, which express the

---

0.20 *the simple and indivisible particles that compose bodies*: the actual constituents from which bodies are built up, and which are not, in turn, composed of other, simpler, particles.

it is likely that we are unacquainted with any elements: Chemistry has no means with which to bring forth the smallest possible particles of any substance; so if that is what "element" means, we have probably never experienced any elements—certainly not knowingly.

0.21 *We cannot be sure that the bodies we regard as simple are not themselves composed...*: Because a substance that has withstood decomposition up to now may someday yield to a hitherto undiscovered technique.

0.22 *I have designated simple substances with simple words*: This principle constitutes the beginning of Lavoisier's effort to *perfect the science* by *perfecting its language* (0.3).

terms ... which ... suggest obviously false ideas: In paragraph (0.33) below, Lavoisier will single out for scorn names like "flowers of zinc" and "butter of arsenic" on this account.

most common and characteristic property of the substances. For beginners, this approach to naming has two advantages. It eases the burden of memorizing names, since it can be difficult for beginners to retain new words that are completely devoid of sense. It also accustoms them from the outset to admit into their vocabularies only those words that are attached to ideas.

0.23 We have designated bodies formed from the combination of many simple substances by names compounded in the very same way. But since even the number of binary compounds is already quite considerable, it was necessary (in order to avoid disorder and confusion) to construct names in such a way as to form classes.

0.24 In the natural order of ideas, the word for a class or genus is the word that calls to mind some property common to a great many individuals: the word for a species, on the other hand, is that which brings to mind the idea of the properties of a few.

0.25 This is not a mere metaphysical distinction as one might suppose: it is, rather, natural. A child, Condillac tells us, will call the first tree we show him "tree." When the child sees a second tree or a third or fourth, he recalls the same idea and gives each of them the same name. In this way, the word "tree," which he first used to refer to an individual, becomes for him a name for a class or a genus, for an abstract idea that comprehends all trees in general. But once we have pointed out to him that all trees do not serve the same purposes, or bear the same fruits, he will soon learn to distinguish them from one another by separate, specific names. This is the logic of every science, and it naturally applies to chemistry.

0.23 *binary compounds*: Substances that are composed of two constituents.

construct names in such a way as to form classes: When one substance is a constituent of several compounds, those compounds form a class. Lavoisier will name the substances of each class in such a way that their names will indicate the constituent they have in common, and thus signal the class to which they belong.

0.24 *genus ... species*: The plural of *genus* is *genera*. Today these terms are most commonly encountered in the system of classification set up by Carl Linnaeus in 1735, in which natural beings were grouped into three kingdoms (animal, vegetable, mineral), each divided into classes, and these in turn into orders, genera, and species, successively. (In some cases, Linnaeus recognized a rank even lower than the species.)

0.25 *This is the logic of every science, and it naturally applies to chemistry*: As Linnaeus named each natural organism according to its genus and species, Lavoisier will form his chemical names—at least for binary compounds—in the same fashion; the following paragraphs offer some fascinating examples.

0.26   Acids, for example, are composed of two substances that we classify as simple. The first constitutes an acid's acidity and is common to all acids, and the name for the class or genus should be derived from this substance. The second is the substance peculiar to each acid, that which differentiates acids from one another, and it is from this second substance that the name of the species should be derived.

0.27   But in most acids, the two constitutive principles, the acidifying principle and the principle acidified, can exist in the proportions that determine their points of equilibrium or saturation. This is what we observe in sulphuric acid and sulphurous acid, and we have expressed these two states of the same acid by varying the ending of the name for the species.

0.28   Metallic substances which have been exposed to the combining powers of air and fire, lose their metallic sheen, gain weight, and take on an earthy appearance. In this state they are composed, like acids, of a single principle common to all of them and a particular principle proper to each. So it was necessary to classify these metallic substances in the same way as acids, by

---

0.26 *Acids*: The French adjective *acide*, which means *tasting sour or bitter*, goes back to 1545. One of the properties, then, which acids have in common is a sour taste. *Vinegar* seems to have been the paradigm of acidic taste; but the herb *sorrel* was another commonly-cited example. Vitriolic acid, associated with sulphur, exhibited vigorous destructive properties and so pointed to other "acidic" properties going beyond taste.

    the first constitutes the acid's acidity: Lavoisier will describe the discovery and naming of this "acid-making" substance in Chapter 4 and its role in the formation of acids in Chapter 5.

    the name for the class or genus should be derived from this substance: Lavoisier's chemical names have a significance that Linnaeus' nomenclature never even aspired to. The common daisy, *bellis perennis*, belongs to the genus *bellis* (from Latin *bellus*, pretty). But no one thinks that every member of that genus contains a unique substance that constitutes its beauty. For acids, however, this *is* the case; moreover, the second of its two constituents is the basis of those properties that distinguish any particular acid from other acids.

0.27 *the two constitutive principles can exist in the proportions that determine their points of equilibrium*: For example, sulphuric acid and sulphurous acid are composed of the same two constituents, but in different proportions. Lavoisier will discuss this variability, and the naming practices that express it, in Chapter 6.

0.28 *Metallic substances ... lose their metallic sheen, gain weight, and take on an earthy appearance*: Here is another class of substances which display similar characteristics. When metals are roasted in air they acquire the properties Lavoisier has listed and become what early chemists called *calces* (singular *calx*). In Chapter 7 Lavoisier will argue that they all share a common constituent—the same one, he claims, that acids do.

first grouping them under a name derived from the principle common to all of them. We adopted the word *oxide* for this purpose. We then differentiated them from one another by names denoting particular kinds of metals.

0.29 The combustible substances that are the specific principles of acids and metallic oxides can be the common principles of a great many other substances as well. For a long time, the only known compounds of this class were compounds of sulphur, but we know today, after the experiments of Vandermonde, Monge, and Berthollet, that charcoal combines with iron and, perhaps, with many other metals, forming steel, plumbago, and so on, depending on the proportion of charcoal. We also know, after the experiments of Pelletier, that phosphorus combines with a great many metallic substances. So we have also classified these different compounds using general names derived from a common substance, with endings that will help remind us of this analogy. We then differentiated them from one another with names derived from the substances proper to each.

* * *

0.32 So at last we have achieved a perspective from which it is possible to recognize a number of different properties of a compound just by looking at its name. We can see which substance in the compound is combustible, whether that combustible substance is combined with an acidifying principle, and, if so, in what proportion. We can also tell at a glance what state an acid is in,

---

we adopted the word oxide: He will replace the old name "calx" with "oxide," a term that derives from the name of the shared constituent.

0.29 *combustible substances*: In common usage, *combustible* materials are those capable of being burnt or consumed by fire. Lavoisier will show that *burning* is a process in which a substance, or one of its constituents, combines with the common principle of acids and oxides.

plumbago: Here, Lavoisier lists it as one of several combinations of charcoal and iron, in keeping with the name *carburet of iron*, which he and his co-authors had introduced earlier in the *Méthode de nomenclature chimique* (Paris, 1787). But "plumbago" was a name far more commonly applied to any of several ores of lead (Latin *plumbum*), or to "black lead," which is not lead at all; see the glossary entries for *plumbago* and *graphite*.

0.32 *to recognize ... properties of a compound just by looking at its name*: The advantages of such names go beyond mere convenience, though the convenience is considerable. In paragraph (0.22) above, Lavoisier recommended that a scientific vocabulary should include "only those words that are attached to ideas." Names like "butter of antimony" are attached to superficial ideas, based on a substance's appearance; but Lavoisier's names are, ideally, based on the actual constitution of each substance and so express knowledge of a type both more profound and more useful.

with which base it is united, whether the saturation is complete, and, if not, whether the acid or base is in excess.

0.33   Clearly, it was not possible to take such different approaches to naming without doing violence to customary usage from time to time, and without adopting terminology that seemed harsh and barbarous at first. But in our experience one quickly develops an ear for new words, especially when they are part of a universal, rational system. Furthermore, the names used up until now, such as "powder of algaroth," "salt of alembroth," "pompholix," "phagedenic water," "turpeth mineral," "colcothar," and many others are no less harsh, no less strange. It requires long practice and a prodigious memory to call to mind the substances these words express, let alone to remember the class of compound to which they belong. The names "oil of tartar per deliquium," "oil of vitriol," "butter of arsenic," "butter of antimony," "flowers of zinc," and so on are still more improper insofar as they give rise to false ideas. For, properly speaking, there are no such things as butters, oils, or flowers in the realm of minerals, much less in the realm of metals, and the substances to which these misleading names refer are, after all, powerful poisons.

0.34   We were criticized when we published our *Essay on Chemical Nomenclature* for having changed the language our teachers spoke, the language that they had explained and passed on to us. But our critics forgot that Bergman and Macquer themselves urged this reform. The learned professor of Uppsala, Bergman, wrote to de Morveau towards the end of his life:

> You must never condone unsuitable terminology: those who are already well-versed in the subject will always understand you, those who are not will learn more quickly.

*   *   *

0.41   I will conclude this preliminary discourse by transcribing word for word some passages from Condillac which seem to me to represent faithfully the

---

with which base it is united: Present-day readers may have been taught to think of *bases* as specifically opposed to *acids*. But Lavoisier uses the term "base" more generically: as the *foundation*, so to speak, of a more complex substance. For example, he regularly speaks of the "base" of a gas—by which he means that substance which, when combined with a sufficient quantity of heat ("caloric"), becomes the gas in question. Elsewhere he writes of "an acidifiable base," meaning a combustible substance which, combining with the universal acid-making substance, forms a specific acid.

     In Chapter 5 he will write: "bases capable of forming neutral salts." While it is indeed *acids* with which such bases must combine in order to form salts, Lavoisier's phraseology presents these bases as the *foundations of salts*, rather than the *antitheses of acids*. For Lavoisier, a "base" is always the base *of* something, not a definite nature in itself.

state of chemistry up until very recently.* These passages, which were written for another purpose, do but gain force when applied to the subject at hand.

> Rather than observing the things that we wished to understand, we preferred to use our imaginations. We became lost in a crowd of errors as we wandered from one false supposition to the next. These errors became prejudices, and so we took them for principles, leading ourselves ever further astray. Because of this, we acquired bad habits of reasoning, and eventually we could only reason in accordance with these habits. For us the art of reasoning became the art of abusing words without properly understanding them… When we have reached such a pass, when errors have piled up to such an extent as this, there remains no other means of retraining the faculty of thought than to forget all that we have known, to take up our ideas again from their origins and follow out their development, to reframe the human understanding, as Bacon puts it.

> The better educated one fashions oneself, the more difficult it will be to begin again. And so scientific works, which are very precise, clear, and well-ordered, will not be accessible to everyone. Those who have never studied will understand them more easily than those who have studied the sciences for a long time, and much more easily than those who have written a lot about them.

0.42   Condillac adds at the end of Chapter 5,

> But the sciences have in fact progressed because natural philosophers have made better observations, and because they articulate them in a language as precise and exact as the experiments themselves. In doing so, they have corrected the language and have reasoned better.

* Condillac, *La logique*, Part II, Chapter 1.

0.41 *forget all that we have known*: Our institutionalized "knowledge" has become so corrupted by unexamined presuppositions and hasty inferences, the only remedy is a radical reconstruction of knowledge, to "take up our ideas again from their origins." These words, written in 1780, can hardly fail to suggest to a modern reader the revolutionary backlash against a corrupt and incorrigible monarchy—a sweeping away that would inflame all France within the decade and take Lavoisier's life in 1794.

Those who have never studied will understand them more easily than those who have studied the sciences for a long time: Readers of this Green Cat Module are urged to take heart from Condillac's words. Bear in mind that Lavoisier is undertaking to free chemical knowledge from the weight of long tradition, as well as from the inherent human tendencies towards facile reasoning. We in the present day live amidst an even greater store of brilliant scientific achievement—which must nevertheless be put aside, at least provisionally, if we wish to experience the foundations for ourselves. If you are a relative novice in the sciences, you will find this task much less onerous than if you had been thoroughly schooled in scientific theory.

Chapter One

On Caloric Compounds and the Formation of Elastic, Aeriform Fluids

1.1   It is an invariable natural phenomenon, and, as Boerhaave has firmly established, a quite general one, that when any body is heated, whether it is a solid or a fluid, it expands in every direction. Facts which may seem to limit the scope of this principle are misleading, or at least are complicated by unusual circumstances that interfere with them. But once we isolate all effects and assign to them their proper causes, we will realize that the separation of particles by heat is a constant and universal law of nature.

1.2   If we first heat a body and, by heating it, increasingly separate its particles from one another, and then we later allow it to cool, these same particles approach one another again, observing the same proportion according to which they separated. The body, at each degree of temperature as it contracts, is restored to the very same dimensions it possessed while it was expanding, and if it is restored to its original temperature, it resumes the same volume it had initially, as far as can be perceived. But since we are still quite far from possessing the means to obtain a degree of absolute cold, and since, so far as we know, there is no degree of coldness that we can assume cannot be increased, we have never yet brought a body's particles as close to one another as possible, which means that the particles of every body in nature do not touch each other. This is a very strange conclusion, but it cannot be denied.

1.3   It is clear that since heat continuously induces the particles of bodies to separate from one another, there are no connections between particles, and there would be no solid bodies at all if the particles were not restrained

---

1.1 *particles*: French *molécules*; but in contrast to the modern technical term "molecules," Lavoisier's "particles" are simply *small portions* of a substance, not necessarily either the smallest possible portions or even of uniform size.

1.2 *the body, at each degree of temperature ... will be restored to the very same dimensions*: A mercury thermometer assumes this principle by treating the volume of the confined fluid as the indicator of temperature.

by another force that tends to draw them together and, as it were, link them to one another. This force, whatever its cause may be, has been called "attraction."

1.4 And so the particles of bodies can be regarded as subject to two forces, one repulsive and one attractive, between which they are in equilibrium. Whenever the attractive force prevails, bodies remain in a solid state, but if, on the contrary, attraction is the weaker force, if heat separates the particles of the body from one another to such an extent that they no longer remain within the spheres of each other's attractive influence, they no longer adhere to one another and the body ceases to be a solid.

1.5 Water constantly presents us with examples of these phenomena. Below zero degrees of the French thermometer, water is in a solid state, and is called "ice." Above this temperature, particles of water are no longer restrained by reciprocal attraction for one another, and they become what is called a "liquid." Finally, above 80 degrees, the particles yield to the repulsive force of the heat. The water takes on the state of a vapor or gas, and converts into an aeriform fluid.

1.6 The same could be said of all bodies in nature. They are either solids, liquids, or in an elastic and aeriform state, according to the ratio that exists between the attractive force of their particles and the repulsive force of heat, or, what comes to the same thing, according to the degree of heat to which they have been exposed.

1.7 It is difficult to conceive of these phenomena without admitting that they are the effects of a real, material substance, an exceedingly subtle fluid that insinuates itself between the particles of bodies and separates them. And even supposing that the existence of this fluid were hypothetical, it explains natural phenomena quite felicitously as will be shown later.

1.8 Since this substance, whatever it might be, is the cause of heat, or, in other words, since the sensation that we call heat is the effect of the accumulation of this substance, we cannot in precise language call the substance "heat." For the same term cannot express both cause and effect. It is this that led me, in

---

1.5 *the French thermometer*: A scale of temperature according to which water freezes at 0° and boils at 80°, as Lavoisier will state explicitly in paragraph 1.13. Today it is commonly called the Réaumur scale.

aeriform fluid: A substance must be either solid, liquid, or aeriform; it is aeriform when it has the air-like state exhibited by gases and vapors. A fluid is anything that flows; thus "fluids" include not only liquids but aeriform substances as well.

a published memoir,* to name it "igneous fluid" and "the matter of heat." But since then, de Morveau, Berthollet, Fourcroy, and I have determined, in our work together on reforming chemical language, that we ought to eliminate periphrases which tend to impede and lengthen discourse, making it less clear and less precise, and which often even suggest ideas that are not sufficiently correct. For this reason we called the cause of heat, i.e., the exceptionally elastic fluid that produces heat, by the name of *caloric*. This expression not only meets the demands of the system we have adopted, it has an additional advantage. It is compatible with all sorts of opinions since, strictly speaking, we are not even obliged to suppose that caloric is real matter. It suffices, as will be better appreciated after reading what follows, that it should be considered a repulsive cause that separates particles of matter, and so the effects of this cause may be treated in an abstract, mathematical way.

1.9   Is light a modification of caloric, or is caloric a modification of light? We cannot possibly decide given the present state of our knowledge. This much is certain, though: in our system we have made it a law to admit only facts and to avoid going beyond the facts whenever possible. So in our system, things should be given different names, provisionally, only if they produce different effects. Thus, we will distinguish light from caloric, but we will admit nonetheless that light and caloric have qualities in common and that in some circumstances they combine in almost the same way and produce some of the same effects.

* *Recueil de l'Académie*, 1777, p. 420.

1.8 *igneous*: Fiery; from Latin *ignis*, fire.

we ought to eliminate periphrases: In classical rhetoric, *periphrasis* or *circumlocution* denotes a roundabout way of speaking. Lavoisier and his collaborators argued that many traditional chemical names consisted of numerous words rather than few because the essential makeup of their corresponding substances was unknown; but that it was now both possible and necessary to replace such names by shorter and more meaningful terms.

caloric: The term *calorique* was originated by Lavoisier; from Latin *calor*, heat.

an abstract, mathematical way: Here, the endeavor to reason about the characteristics of an effect without making specific assumptions concerning its cause. Lavoisier will reflect again on such reasoning in paragraph 1.45 below.

1.9 *Is light a modification of caloric?*: That is, is light a *form* of caloric, rather than an essentially different entity? Since heat and light are so often found together, it is reasonable to ask whether one is just a special form of the other.

1.10   The foregoing already suffices well enough to fix the idea that ought to be attached to the word "caloric." But I must still perform a more difficult task: I must find a way to convey proper ideas of the manner in which caloric acts on bodies. Since this subtle matter penetrates all of the interstices of all substances we know, and since there are no vessels from which it cannot escape, which means that nothing can contain it without some loss, it is only possible to understand its properties through its effects which, for the most part, are fleeting and hard to grasp. In the case of things that can be neither seen nor felt, it is of the utmost importance to guard against flights of the imagination, since the imagination always tends to overleap the truth, and can hardly be confined within the limits circumscribed by facts.

1.11   We have just seen that the same body becomes a solid, a liquid, or an aeriform fluid according to the quantity of caloric that permeates it; or, to put this more rigorously, according to whether the repulsive force of the caloric is equal to, stronger than, or weaker than the force of attraction among the particles.

1.12   But if only these two forces existed, there would be no range of temperature at which bodies would be liquids, and they would pass suddenly from a solid state to a state of elastic, aeriform fluid. Thus water, for instance, would begin to boil at the moment it ceased to be ice. It would convert into an aeriform fluid, and its particles would disperse throughout space indefinitely. That this does not happen is because a third force, the pressure of the atmosphere, inhibits this dispersion. It is because of this pressure that

---

1.10 *to fix the idea*: That is, to determine or specify the idea.

1.12 *these two forces*: That is, the attractive force of particles of the substance for one another, and the repulsive force exerted by caloric as it insinuates itself between those particles.

if only these two forces existed, there would be no ... liquids: Since liquids *flow*, they must contain sufficient caloric to have overcome the mutual attraction of their particles. But if their particles are not attracting one another, what keeps them from dispersing indefinitely throughout space? Yet liquids have a definite volume. Thus without some third force capable of holding the particles together, *liquids* would be impossible.

the pressure of the atmosphere: Torricelli filled a long tube with mercury, then inverted its open end into a basin of mercury; the mercury column fell to a height of about 30 inches, regardless of the length of the tube—and therefore no matter how tall the column had been originally. *Something* must be balancing the weight of the column; he argued that only the pressure of the atmosphere upon the surface of the mercury reservoir could plausibly perform that function.

30 in.

water remains in a liquid state between 0 and 80 degrees of the French thermometer. The quantity of caloric that water absorbs at these temperatures is insufficient to overcome the pressure exerted by the atmosphere.

1.13   One can see then that without the pressure of the atmosphere, there would be no stable liquids. We would see bodies in this state only at the very point of melting, for if they were to absorb even the slightest additional heat their parts would separate from each other and disperse. What is more, without the pressure of the atmosphere we could not even properly speak of aeriform fluids. For at the very moment at which the repulsive force of caloric prevailed over the force of attraction, the particles would fly away from each other and disperse indefinitely unless their own weight were to bring them back together so as to form an atmosphere.

1.14   A few simple reflections on well-known experiments should be enough to help us see the truth of what I have just claimed. And this is also confirmed in a straightforward way by the following experiment, which I have already explained in detail to the Academy in 1777.*

1.15   A narrow glass vessel, A, set on its base, P, is filled with sulphuric ether (Plate VII, Figure 17).** This vessel should have a diameter of no more than 12 or 15 lines and should be about 2 inches tall. It should be covered with a moistened bladder which has been fastened to it very tightly by a thick wire wrapped around its neck many times. For greater security, a second bladder should be placed over the first and fastened in the same manner. The vessel should be completely filled with ether so that no air remains between the liquid and the bladder.

Fig. 17.

* *Recueil de l'Académie,* 1777, p. 426.

** Later I will give a definition for the liquid that is called "ether," and I will explain its properties. I will only say for now that "ether" refers to a highly volatile, inflammable liquid with a specific weight much less than that of water, even less, in fact, than spirit of wine.

1.15 *sulphuric ether*: The liquid produced by distilling alcohol ("spirit of wine") with sulphuric acid. As Lavoisier states in his footnote, it is both volatile and flammable.

12 or 15 lines: the *line* (French *ligne*) is a unit of length equal to .089 inch. See the table of French pre-revolutionary weights and measures in the Appendix.

a moistened bladder: Animal bladders are highly elastic and can be inflated like modern rubber balloons. Pigs' bladders, especially, were used for pretty much the same purposes as today's balloons—both utilitarian and entertaining. Latex balloons were produced in the 1840's but were not widely manufactured until the 1930's.

It is then placed beneath the receiver, BCD, of an air pump, the top of which has been covered with a leather case through which passes a rod, EF, with a very sharp blade or point at the bottom, F. A barometer, GH, should be attached to the same receiver.

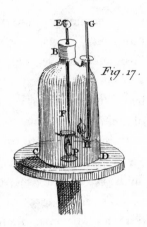

Fig. 17.

1.16   Once all of this has been arranged, the chamber is evacuated. Then the pointed rod, EF, is lowered until it punctures the bladder. Instantly, the ether begins to boil away with astonishing rapidity. It evaporates, and is transformed into an elastic, aeriform fluid, which completely fills the receiver. If there is enough ether that some drops of it remain in the vial after the evaporation has stopped, the elastic fluid produced can support, in winter, an 8 to 10 inch column of mercury in the barometer attached to the air pump, or a 20 to 25 inch column in the heat of summer. It is possible, in order to make the experiment more complete, to introduce a little thermometer into the vessel A which

---

1.15, cont. *the receiver, BCD, of an air-pump*: An odd name, one might think, for this vessel since, so far from "receiving" anything, it will have air pumped *out of* it. But containers of this size and style had long been used for collecting the products of distillation, and so had acquired the name "receiver" with good reason. Although the utensil later found other uses, it was still *crafted* in the same way and was therefore, as far as its makers were concerned, still a "receiver." In 1660 Robert Boyle described a container "which we, with the Glass-men, shall often call a Receiver, for its affinity to the large Vessels of that name, used by Chymists" (Oxford English Dictionary).

leather case: This forms an airtight seal about rod EF, while allowing it to move up and down.

a barometer, GH: A mercury-filled tube, sealed at the top as shown here (compare the sketch accompanying the comment to 1.12).

1.16 *the elastic fluid ... can support ... an 8 to 10 inch column of mercury*: When the apparatus is cold (in winter), the aeriform ether exerts sufficient pressure within the receiver to support the weight of 8–10 inches of mercury. But why is the pressure so much greater in summer, as he goes on to report? Since accumulation of caloric is the cause of the sensation of heat (1.8), a higher temperature indicates the presence of a greater quantity of caloric. And since caloric is itself preeminently elastic, a greater quantity confined in the same space will exert greater pressure. An accumulation of caloric will also cause more liquid to evaporate, as Lavoisier outlined in paragraph 1.11.

contains the ether. The temperature drops considerably while the ether is evaporating.

1.17 In this experiment the weight of the atmosphere, which ordinarily presses on the surface of the ether, is eliminated, and that is all. The effects that follow from this clearly prove two things. First, at the temperatures in which we live, ether would always remain in the state of an aeriform fluid if it were not for the pressure of the atmosphere. Second, the passage from the liquid state to the aeriform state is accompanied by considerable cooling due to the fact that some of the caloric, either free or in equilibrium with the surrounding bodies, combines with the ether during vaporization in order to convert it into an aeriform fluid.

1.18 This same experiment succeeds with all evaporable fluids, such as spirit of wine (i.e., alcohol), water, or even mercury, but with this difference: the atmosphere of alcohol that forms under the receiver supports, in winter, a column of mercury of only one inch above the level of the barometer attached to the air pump, and four or five inches in summer. Water can only support a column of a few lines, mercury a fraction of a line. Less fluid evaporates when working with alcohol than when working with ether, even less in the case of water, less

---

the temperature drops considerably while the ether is evaporating: A liquid evaporates (becomes aeriform) by acquiring additional caloric (1.11). This caloric can only be supplied by the surrounding bodies—especially by the remaining liquid. As that residue loses caloric, its temperature drops.

1.17 *the weight of the atmosphere ... is eliminated*: Before the receiver was evacuated, the surface of the liquid ether in vessel A was pressed by the full weight of the atmosphere; and that pressure was maintained by the bladders tied over A, even after the air was pumped out of the receiver. But the instant the bladders were punctured, the liquid surface became exposed to an environment from which the atmosphere's weight had been eliminated. As Lavoisier points out, elimination of atmospheric pressure is all that was necessary to allow the liquid to become aeriform.

caloric ... combines with the ether: Notice that all of the phenomena discussed so far appear to involve caloric that is somehow attached to, or combined with, other substances. Caloric can, evidently, pass from one substance to another—as when a hot body warms a cold one by contact—but we have not seen any indications that caloric can accumulate *on its own*, apart from another material. Why would caloric always have to be united with some other substance? And what might be the nature of such combination? Lavoisier will say more about the relations between caloric and other materials in paragraph 1.36.

1.18 *the same experiment succeeds with all evaporable fluids*: But he finds that each liquid develops its own characteristic pressure in the aeriform state. When they are at the same temperature, *ether, alcohol, water*, and *mercury* exhibit aeriform pressure in descending order.

still in the case of mercury. Because of this, less caloric is involved and so there is less cooling, and this agrees perfectly with the results of experiments.

1.19   Another kind of experiment proves just as clearly that the aeriform state is a modification of bodies and that it depends on the temperature and pressure to which they are exposed.

1.20   In a memoir that we read to the Academy in 1777 but have yet to publish, Laplace and I showed that when ether was submitted to a pressure of 28 inches of mercury, that is to say, to a pressure equal to that of the atmosphere, it began to boil at 32 or 33 degrees of a mercury thermometer. De Luc has made analogous experiments with spirit of wine and has established that it begins to boil at 67 degrees. And everyone knows that water begins to boil at 80 degrees. Since boiling is nothing but the evaporation of a fluid, or the point at which it passes from a liquid to an elastic, aeriform state, it was clear that when ether is kept at a temperature higher than 33 degrees and at an ordinary degree of atmospheric pressure, it ought to exist in the state of an aeriform fluid, and that the same thing ought to happen for spirit of wine above 67 degrees and water above 80. All of this is completely confirmed by the following experiments.*

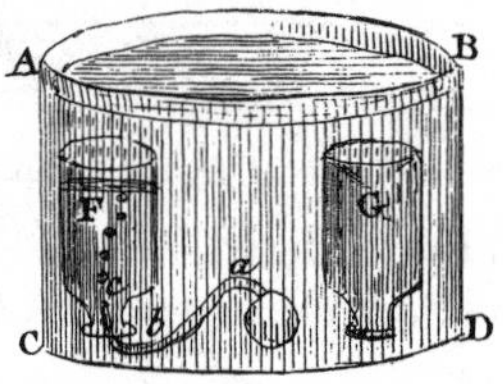

1.21   I filled a large vessel, ABCD, with water at a temperature of 35 or 36 degrees (see Plate VII, Figure 15). In order to more easily appreciate what is happening inside, I assume that the flask is transparent. It is possible to immerse one's hands for a long time without discomfort in water at this temperature. I submerged bottles F and G upside down in the jar, first filling them and then upending them so that they stood on their necks at the bottom.

Fig. 15.

1.22   This being done, I filled a very small long-necked flask, the neck *abc* of which had been bent twice, with suphuric ether. I submerged this flask in vessel ABCD, and managed to introduce the neck of the flask under the neck of one of the bottles, F, in the manner illustrated in the figure. As soon as it

---

* *Mémoires de l'Académie*, 1780, p. 335.

1.19 *the aeriform state is a modification of bodies*: In becoming aeriform, a body merely takes on a different form; it does not change its essential nature or become a different substance. Recall Lavoisier's similar use of "modification" in paragraph 1.9.

1.20 *28 inches of mercury ... equal to the pressure of the atmosphere*: Lavoisier's "inch" is actually the *pouce*, equal to 1.066 modern inches (see the table of French pre-revolutionary weights and measures in the Appendix). So his determination of atmospheric pressure would actually be 29.8, expressed in modern inches.

began to feel the effects of the heat, the ether began to boil, and the caloric that combined with it converted it into an elastic, aeriform fluid, with which I filled several bottles like F and G.

1.23   Here is not the place to examine the nature and properties of this aeriform fluid, which is extremely flammable; but without relying prematurely on knowledge I ought not to assume the reader possesses, and focusing only on the matter before us at the moment, I will observe, on the basis of this experiment, that it is almost the case that ether can only exist in an aeriform state on the planet we inhabit. If the pressure of our atmosphere were equivalent to a column of 20 or 24 inches of mercury rather than 28, we could not obtain ether in a liquid state, at least during the summer. Accordingly, it would be impossible to find liquid ether on mountains at higher elevations, for it would convert to gas as soon as it formed unless very strong flasks were used to condense it and it was both pressurized and cooled. Finally, since the temperature of blood is just about the temperature at which ether passes from a liquid to an aeriform state, ether ought to evaporate as soon as it is introduced into the body; and it is quite likely that the medicinal properties of ether depend on this, so to speak, mechanical effect.

1.24   These experiments succeed even better with nitrous ether, because it evaporates at a lower temperature than sulphuric ether. As for alcohol or spirit of wine, the experiment required to obtain it in an aeriform state is somewhat more difficult, because this fluid only evaporates at 67 degrees Réaumur. So it is necessary to submerge the flask of alcohol in water that is almost boiling, at which temperature the water is too hot to place one's hands in it.

1.25   It was clear that the same ought to hold for water, that this fluid should also convert into gas when exposed to greater heat than is required to make

---

1.22 *sulphuric ether*: The same liquid used in the previous experiment (1.15).

1.23 *the medicinal properties of ether*: If Lavoisier means its effects as a general anesthetic, his conjecture is probably erroneous. The physiological processes associated with anesthesia are only imperfectly understood; but they are unlikely to be "mechanical," as Lavoisier suggests, in any significant degree.

1.24 *nitrous ether*: It is prepared in the same way as sulphuric ether, but using nitric instead of sulphuric acid. Lavoisier will discuss both these acids in Chapter Six.

degrees Réaumur: Degrees of "the French thermometer" (1.5) originally introduced by René Antoine Ferchault de Réaumur in 1730.

1.25 *gas*: Lavoisier previously (1.5) offered *gases* and *vapors* as examples of aeriform fluids. Some chemists regarded gases and vapors as different in kind, but for Lavoisier there is no distinction. He will adopt "gas" as the general term in paragraph 1.29.

it boil. But although we were convinced that this was true, Laplace and I thought we ought to confirm it with a direct experiment, the result of which was as follows. We filled a glass jar, A, with mercury, turned it upside down, and placed it in a mercury-filled dish, B, which we slid beneath it (see Plate VII, Figure 5). Into the jar we introduced about 2 gros of water, which stood at height CD and sat above the surface of the mercury. Then we submerged all of this in a large iron boiler, EFGH, which had been placed on top of a furnace, GHIK. This boiler was full of boiling brine, the temperature of which was above 85 degrees. As we know, water rich in salts can be raised to a greater degree of heat than boiling water.

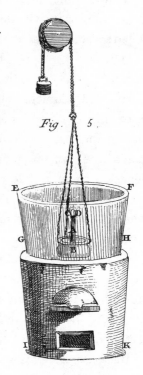

Fig   *5*

1.26   As soon as the 2 gros of water placed in the upper part, CD, of the jar or tube reached a temperature of about 80 degrees it began to boil, and instead of occupying the small space ACD, it was converted into an aeriform fluid that filled the jar completely. The mercury even descended a little below its ordinary level, and the jar would have flipped over if it had not been so thick and, for this reason, very heavy, and if it were not also attached to the dish by iron wire. Moments after we removed the jar from its salt-water bath, the water condensed and the mercury rose again. But the water resumed its aeriform state as soon as the assembly was submerged again.

1.27   And so there are quite a few substances which convert into aeriform fluids at temperatures very close to those in which we live. We will see shortly that

---

1.25, cont. *2 gros of water*: About $7\frac{1}{2}$ grams or $\frac{1}{4}$ US ounce. See the table of French pre-revolutionary weights and measures in the Appendix.

water ... stood at height CD: A bent tube like that used in (1.22) would serve to bring water into jar A from the bottom. The water would rise to the top at CD, displacing some of the mercury that formerly filled the jar.

salts: Originally, "salts" denoted substances that resembled common salt in appearance or in properties such as solubility. In later chapters (not included in the present selection), Lavoisier will offer a more systematic definition of salts and classify them as *acid, alkaline,* or *neutral.*

raised to a greater degree of heat than boiling water: A boiling liquid does not increase in temperature no matter how intensely it may be heated; it merely boils more furiously and vaporizes at a greater rate. That stability is the reason why the boiling

there are others such as marine or muriatic acid, volatile alkali (ammoniac), carbonic acid (fixed air), sulphurous acid, and so on, which always remain in an aeriform state under the ordinary conditions of heat and pressure of the atmosphere.

1.28   All of these facts, of which I could easily give many additional examples, entitle me to assert as a general principle what I have already said above: that almost all bodies in nature can exist in three different states, in a solid state, a liquid state, and an aeriform state, and that these three states of the same body depend on the amount of caloric with which it is combined. From now on, I will designate aeriform fluids by the general term "gas," and so for each kind of gas I will say that caloric, which assumes in some way the role of solvent in every case, should be distinguished from the substance with which it combines and which forms the gas's base.

---

point of water is a useful standard for thermometry. But water containing dissolved substances generally boils at a higher temperature than does pure water; Lavoisier's brine (sea-water) boiled at 85° Réaumur; pure water would have boiled at 80°.

1.28 *the role of solvent*: When one substance dissolves in another, the ingredient that dissolves is the *solute*; the substance in which it dissolves is the *solvent*. When, for example, salt dissolves in water, salt is the solute, water is the solvent.

the substance with which it combines and which forms the gas's base: The *base* of a gas is that solid or liquid substance which, by combining with sufficient caloric, takes on the gaseous form. Water, for example, is the base of steam. In Lavoisier's image of the gaseous state as a *solution*, if caloric is the solvent, the base is the solute.

Lavoisier generally uses the term "base" to mean something like *the foundation upon which a substance is formed*. The "base" of a substance is, at least generally, the base *of that substance only*—or at most a few others.

Note: Readers who have had any exposure to chemistry, or even to gardening, will have become familiar with a modern sense of *base* as the opposite of *acid*. Although the modern usage derives to some extent from Lavoisier's, his notion of a "base" is far more general. For Lavoisier, *every acid begins with its base*, which is thus a constituent of the acid rather than a counterpart to it.

Notice also that Lavoisier states that the base of a gas *combines* with caloric, even though he is comparing gases to *solutions*. Today we distinguish sharply between *solutions*—in which the same quantity of solvent can take up a continuously varying amount of solute—and *true compounds*, which are characterized by *fixed proportions* of the combining substances. Though Lavoisier is exquisitely aware that many substances combine in definite ratios, he has no way, as yet, to discover *how* substances associate with one another; so for him it is far from obvious that the distinction between definite ratios and varying ratios of combination is a fundamental one. He will, in fact, pay attention almost exclusively to fixed-ratio combinations, but probably not for any strict theoretical reasons.

1.29   We have been obliged to give names to these bases of gases, bases which are still not widely known. I shall indicate them in Chapter Four of this work after I have given an account of some of the phenomena associated with heating and cooling bodies and have supplied more precise ideas about the constitution of our atmosphere.

1.30   We have seen that the particles of all bodies in nature are in a state of equilibrium between attraction, which tends to bring them together and unite them, and the influence of caloric, which tends to separate them. This implies that caloric not only surrounds every part of every body, it also fills the interstices between their particles.

1.31   It is possible to form an idea of these tendencies if one imagines a vessel filled with little balls of lead into which a very fine powder, like sand, has been poured. In this case, it is clear that the sand will spread out uniformly and fill the spaces left between the balls. The balls in this example are to the sand as the particles of bodies are to caloric, except that the balls in the example touch each other while the particles do not, but rather maintain a little distance from one another due to the influence the caloric exerts.

1.32   If one were to substitute hexahedrons or octahedrons or some other regular bodies of equal solidity for the round balls, there would no longer be the same amount of space between the objects in the jar and it would no longer be possible to fill that space with the same amount of sand. It is the same for all bodies in nature. The space between the particles of different bodies is not of equal volume, and the volume depends on the shape of the particles, their size, and on the distance from each other at which they are maintained according to the ratio between their force of attraction and the repulsive force exerted by the caloric.

1.33   It is in this sense that the phrase "the capacity of bodies for containing the matter of heat" should be understood, which is a particularly apt expression

---

1.32 *hexahedrons or octahedrons*: Here, six- or eight-sided solid bodies.

no longer be the same amount of space: Since bodies having flat sides would not pack together the same as round balls, the amount of space between them would also not be the same as the space between round balls.

1.33 *the capacity of bodies for containing ... heat*: Joseph Black found that when equal weights of hot mercury and cool water were stirred together, the mixture attained a common temperature much closer to the initial temperature of the water than to that of the mercury. But the heat lost by the mercury must have been gained by the

introduced by the English natural philosophers who were the first to hold precise notions about this subject. An example of what happens in the case of water, and several reflections on the manner in which water saturates and penetrates bodies, will make all of this more intelligible. Clearer understanding of abstract things is always possible with the aid of sensible comparisons.

1.34   If you submerge pieces of wood of different varieties of the same volume (say a cubic foot) in water, the fluid will seep into the pores bit by bit. The pieces of wood will swell and gain weight, but each piece will admit a different amount of water into its pores. The lightest and most porous pieces will take in more, the heavier and denser pieces only a very little. Finally, the proportion of water that they soak up will also depend on the nature of the particles that constitute the wood—on the greater or lesser affinity that they have for water; and very resinous wood, for example, despite being very porous admits very little. So one could say that different kinds of wood have different capacities for receiving water, and it is even possible to know how much water a piece of wood has absorbed by considering how much weight it has gained. Since, however, we do not know how much water it contained before it was immersed, it will be impossible to know the absolute quantity it contains in the end.

1.35   The same considerations apply to bodies immersed in caloric, though it should be observed that water is an incompressible fluid, while caloric is

---

water; so equal quantities of heat gained or lost resulted a *small* temperature change for water, but a *large* temperature change for mercury.

Now if you pour a few ounces of water into a drinking glass, the water level might rise by several inches. But the same amount of water poured into a capacious mixing bowl will produce a much smaller rise. Similarly, Black interpreted a body's change in temperature when it gained or lost heat as an indication of that body's *capacity for heat*: the smaller the temperature change (for a given quantity of heat), the greater its capacity.

the matter of heat: This expression was widely employed to indicate the treatment of heat as a substance. Lavoisier employed it in a 1777 memoir but, for the reasons he noted in paragraph 1.8, subsequently adopted the term "caloric."

the English natural philosophers: Prominent among these in the eighteenth century would be Joseph Black, Joseph Priestley, and John Dalton. Among them, however, it is probably Black who may best be said to have held "precise notions" concerning the heat capacity of bodies.

endowed with great elasticity, that is to say, in other words, that the particles of caloric have a strong tendency to separate from each other when some force draws them together. It is clear that this factor must lead to very notable differences in the results.

1.36   Now that we have brought these issues to such a point of clarity and simplicity, it will be easy for me to expound the ideas that ought to be attached to the expressions "free caloric," "combined caloric," the "specific quantity of caloric" contained in different bodies, "capacity to contain caloric," "latent heat," and "sensible heat." These expressions are by no means synonymous: each has a strict and determinate sense, in light of what I have just explained. So I shall attempt to fix the senses of these terms with the following definitions.

1.37   "Free caloric" is caloric which is not involved in any combination with any other body. But since we live amidst a system of bodies to which caloric adheres, it so happens that we can never obtain this principle in an absolutely free state.

1.38   "Combined caloric" is caloric which is locked up within bodies by the force of affinity or attraction, and which constitutes an aspect of the substance, and even the solidity of bodies.

1.39   By the expression "specific caloric" of bodies is meant the quantity of caloric required to raise the temperature of bodies, equal in weight, the same number of degrees. This quantity of caloric depends upon the distance between the particles of bodies—upon their lesser or greater adherence; and it is this distance, or rather

---

1.35 the *particles of caloric have a strong tendency to separate from one another when some force draws them together*: That is, the particles of caloric *resist* any effort to squeeze them together. This is caloric's "great elasticity," just as a foam rubber ball resists the effort to compress it and regains its original volume as soon as the compression is removed. Lavoisier will make this tendency to separate a *definition* of elasticity in paragraph 1.45 below.

1.37 *this principle*: That is, *caloric*. For Lavoisier, a "principle" is primarily a *rule* or *standard*—as, for example, the "principle" that *heated bodies expand*, which he cited in paragraph 1.1. But he also uses the term to denote a substance that, so far as we can tell, is *simple* and not a composition of other substances. Such is its usage here. Recall that in his Preliminary Discourse, paragraph 0.20, he treated "principles" and "elements" as virtually synonymous.

the space that results from it, which (as I have already noted) has been named "the capacity to contain caloric."

1.40   "Heat," considered as a sensation, or, in other words, "sensible heat," is nothing else but the effect produced on our organs of sense by the caloric emanating from surrounding bodies.

1.41 In general, we only experience sensations when there is movement. We can put this down as an axiom: "no movement, no sensation." This general principle naturally applies to the sensation of cold and heat: when we touch a cold body, the caloric, which seeks equilibrium in all bodies, passes from our hand into the body we are touching, and we feel the sensation of cold. The opposite effect occurs when we touch a hot body: caloric passes from the body into our hand and we experience a sensation of heat. If the body and our hand are the same or approximately the same temperature, we feel nothing, neither cold nor heat, for in this case there is no movement, no caloric is transmitted, and so once again there is no sensation without a movement to cause it.

1.42   When the thermometer rises, it is a certain indication that caloric is spreading into the surrounding bodies. The thermometer is just one of these bodies, and it receives its portion of caloric in accordance with its capacity for containing caloric and its mass. The thermometer only registers a change in the system of bodies to which it belongs, a displacement of caloric. It indicates nothing else but the amount of caloric it has received and does not measure the total quantity disengaged, displaced, or absorbed.

1.39 *the space that results ... has been named "the capacity to contain caloric"*: Imagine equal volumes of two solids or liquids A and B; and suppose A has more space between its particles than does B. Then equal amounts of caloric, added to both, will be *more highly compressed* in B and must then exert greater repulsive force in B than in A. But Lavoisier has implicitly associated a body's expansion, and hence its temperature, with this repulsion (1.2, 1.7, 1.12). Adding equal amounts of caloric to both bodies, then, would produce a greater temperature change in B than in A. Black (1.33, comment) would therefore have declared A to have *greater capacity for heat* than B; and Lavoisier is pointing out that the substance that has greater capacity for heat is also the substance that has *more space between its particles*.

1.42 *The thermometer ... indicates nothing else but the amount of caloric it has received*: Even though we commonly speak of the thermometer as measuring the temperature of a body to which it is applied, or of a liquid in which it is immersed, it properly should be said to indicate *its own* temperature—for the length of its mercury column expands and contracts according to the amount of caloric it has received from the body or liquid being measured.

1.43   Laplace invented the simplest and most precise way to measure the quantity of caloric disengaged. This method is described in a memoir,* and a summarized explanation of the procedure appears at the end of this work. It involves placing a body or several bodies releasing caloric in the middle of a hollow sphere of ice. The quantity of ice melted is an exact expression of the amount of caloric released.

* * *

1.44   Before concluding this chapter, it only remains for me to say a word about the cause of the elasticity of gases and vaporous fluids. It is not hard to see that this elasticity is related to the elasticity of caloric, which seems to be the most elastic body in nature. Nothing is simpler than to imagine that a body becomes elastic through its combination with another body which is itself endowed with that property. But admittedly this would be to explain elasticity by elasticity, which merely shifts the problem back a step and leaves us, as before, wondering what elasticity truly is, and why caloric is elastic. Considered in the abstract, elasticity is simply the property that bodies have of separating from one another when they are forced together. This tendency that particles of caloric have—that of separating themselves—extends even to very great distances. To be convinced of this, consider the fact that the air is susceptible to very high compression, and this means that particles of air are already very distant from one another since the distance between them

* Mémoires de l'Académie, 1780, p. 364

1.43 *a summarized explanation of the procedure appears at the end of this work*: The device used to measure the quantity of caloric that different substances release is the *ice calorimeter*. Lavoisier's summary is not part of the present selections; but, briefly, the procedure is this: Ice fills the space *bb* between the outer compartment and the inner chamber A. The body under test is placed in A, and the apparatus covered with an insulating lid. The test body communicates caloric to the ice, melting it; and the amount of water that drips from spout *y* is a measure of the amount of caloric released.

1.44 *elasticity is simply the property...*: Lavoisier noted the tendency of particles to separate, and so resist being pressed together, in paragraph 1.36 above. Now he identifies that property as the *definition* of elasticity.

must be at least sufficient for them to get so much closer to one another. Now these particles of air, which are already quite distant from each other, tend to separate even further: indeed, if Boyle's vacuum is created in a large receiver, the last portion of air that remains spreads out uniformly throughout the whole volume of the vessel. No matter how large the vessel may be, the air fills it entirely and presses against its sides. Now this effect can only be explained by assuming that the particles endeavor to spread out in all directions, and it is unknown whether there is a distance beyond which this phenomenon no longer occurs.

1.45   There is, then, a real repulsion between the particles of elastic fluids, or at least everything occurs in the same way it would if there were such a repulsion. It seems permissible, then, to conclude that the particles of caloric repel one another. Once we admit this force of repulsion, explaining the formation of aeriform fluids or gases becomes very simple. But at the same time it must be admitted that a repulsive force which acts between tiny particles at great distances from one another is hard to conceive.

Boyle's vacuum: Robert Boyle's improved air-pump allowed him to evacuate the air from a large "receiver," and by such means to demonstrate that air was necessary for the transmission of sound, for the combustion of a candle, and for the maintenance of life.

1.45 *at least everything occurs in the same way it would if...*: Since there is no way to observe the particles of elastic fluids directly, Lavoisier cannot positively affirm that they exert repulsive forces on one another. Their mutual repulsion must remain a hypothesis; but since (as Lavoisier has shown) it would explain so many things if it were true, the hypothesis is a very useful one.

To reason freely on the basis of this hypothesis—drawing conclusions from it without necessarily claiming certainty for it—is an example of treating phenomena "in an abstract, mathematical way," as Lavoisier expressed it in paragraph 1.8. Evidently, for him, a *mathematical* way need not involve symbols or equations. What more inclusive sense of "mathematical" might he have in mind?

a repulsive force which acts ... at great distances ... is hard to conceive: One might regard *any* force acting at great distances, whether repulsive or attractive, as "hard to conceive"; Newton had to face such an objection to his theory of universal gravitational attraction. Is there any reason why a *repulsive* force would be particularly challenging?

1.46   It may seem, perhaps, more natural to suppose that the particles of caloric attract one another more than the particles of bodies do, and that the particles of bodies only separate from one another because of the attractive force that causes particles of caloric to unite. Something analogous to this phenomenon happens when a dry sponge is submerged in water: it swells, and its particles separate from one another as water fills the space between them. It is clear that the sponge, in swelling, has acquired a greater capacity to contain water than it previously had. But should we say, in this case, that the water communicates a repulsive force to the particles of the sponge that tends to separate them from one another? Certainly not: in this case only attractive forces are at work. These forces are, first of all, the weight of the water and the effort that it exerts in all directions like any fluid; second, the attractive force between the particles of water; third, the attractive force between the particles of the sponge, and finally the reciprocal attraction between the particles of the water and the sponge. It is easy to understand that the explanation of the phenomenon depends upon the intensity and relation of all these forces. It is likely that the separation of the particles of bodies by caloric depends in the same way on some combination of different

----

1.46 *more natural to suppose that the particles of caloric attract one another more than the particles of bodies do*: On this alternative there is no need to hypothesize *repulsion* at all: if *all* particles—those of caloric as well as those of bodies—*attract* rather than repel one another, then in their competition to congregate, whichever particles have less mutual attraction will be driven apart as those with stronger attraction succeed in coming together.

But why would anyone find this alternative supposition "more natural"? Some possible reasons: (i) the alternative hypothesis is simpler because it posits only one force (attraction) rather than two (attraction and repulsion); (ii) Newton's gravitation theory has already given us good reason to suppose that all particles attract one another; (iii) Lavoisier's example of the sponge, to which he devotes the rest of the paragraph, helps make the alternative hypothesis visualizable, and so escapes the objection "hard to conceive" in paragraph 1.45.

Notice how detailed the foregoing speculation has become! Since in fact we have no access to the individual particles of bodies, is such meticulous conjecture really compatible with the "abstract, mathematical way" (1.8; 1.45, comment)? Lavoisier will end this paragraph with what appears to be a corrective effort.

attractive forces. It is the result of these forces that we attempt to express, in a way that is more concise and better conforms to the imperfect state of our understanding, by saying that caloric communicates a repulsive force to the particles of bodies.

---

express in a way that ... better conforms to the imperfect state of our understanding: Lavoisier implies a profound maxim here. It is better to characterize things in a general and imprecise way—if such indeed is the state of our knowledge about them—than to weave elaborate and detailed hypotheses that misleadingly imply a level of precision that our thinking does not actually possess. Some eighty years after the publication of Lavoisier's *Traité*, James Clerk Maxwell in his own *Treatise* would follow an equivalent maxim when propounding a new mathematical science of electricity and magnetism.

caloric communicates a repulsive force to the particles of bodies: Lavoisier reasserts his original supposition, which represents the phenomena in a conveniently concise way, and without *explaining too much*.

Chapter Two

Overall Views on the Formation and Constitution
of the Earth's Atmosphere

2.1   The account of the formation of aeriform elastic fluids, or gases, that I have just presented sheds great light on the way in which the atmospheres of planets, in particular the earth's, were formed in the beginning. It is clear that our atmosphere must be a product and an amalgam of (1) all evaporable substances or, better, all substances which remain in an aeriform state at the temperatures in which we live, and at a pressure equal to a column of mercury 28 inches high; and of (2) all fluids or solids that can be dissolved in this assortment of different gases.

2.2   To better fix our ideas concerning a matter which no one has yet considered carefully enough, let us imagine for a moment what would happen to the various substances which compose the earth if the temperature were to change abruptly. Let us imagine, for example, that the earth were suddenly transported to a much hotter region of the solar system, perhaps somewhere near Mercury, where the normal temperature is probably much higher than that of boiling water. Before long, water, all fluids which evaporate at any temperatures close to that of boiling water, and mercury itself, would begin to expand. They would all convert into aeriform fluids or gases, and would become part of the atmosphere. These new types of air would mix with those already present, resulting in reciprocal decompositions and new compounds, until the principles composing these different airs or gases should achieve a state of equilibrium determined by their respective affinities for one another.

2.2 *new types of air*: Although "air" most often refers to the gaseous substance that surrounds the earth and which animals breathe, it also has a more general meaning, as here, denoting any gas or vapor.

2.3   But we should not lose sight of the fact that this vaporization would be limited. For as the quantity of elastic fluid increased, the weight of the atmosphere would increase proportionally. And since any pressure would pose an obstacle to vaporization, and since the most evaporable fluids can resist evaporating under great heat so long as that heat is counteracted by a proportionally greater pressure, and, finally, since all liquids, even water itself, can sustain a red heat in Papin's machine, it is clear that the new atmosphere would reach such a degree of heaviness that any water which had not yet evaporated would cease boiling and would remain in a liquid state. As a result, even in the case we are imagining, or in any other case of the same kind, the weight of the atmosphere would be limited and could not surpass a certain boundary. We could carry these reflections much further and examine what would happen to stones, to salts, and to many other substances which can be melted and compose the earth. We could imagine how they would soften, begin melting, and become fluids. But these are digressions, and I hasten to return to the point.

2.4   If the earth were suddenly placed, instead, in some very cold region, just the opposite would occur. The water that makes up our rivers and oceans, and probably most known liquids, would be converted into solid mountains, and into extremely hard rocks. These would be translucent, homogeneous, and white at first, just like rock crystal. But over time, by mixing with substances of different natures, they would become diversely colored opaque stones.

2.3 *the weight of the atmosphere would increase*: The substances which form earth's atmosphere have weight, even though they are in an aeriform state. As more and more materials take on a gaseous condition, so much greater will be the total weight of those gases composing the atmosphere.

any pressure would pose an obstacle to vaporization: Recall Lavoisier's argument that the pressure of the atmosphere is what prevents immediate vaporization of solids and so makes the liquid state possible (1.13).

water ... can sustain a red heat in Papin's machine: Also known as Papin's digester, pictured here, this early pressure cooker was capable of developing pressures so high that water remained liquid even when the apparatus was heated to a red glow.

2.5   Under this supposition the air, or at least some of the aeriform substances composing it, would undoubtedly cease to exist as elastic vapors, due to an insufficient degree of heat. They would therefore return to a liquid state, and we have no idea of the new liquids they would form.

2.6   These two radical suppositions make it clear that (1) *solidity, liquidity,* and *elasticity* are different states of the same matter: three specific modifications through which nearly all substances can pass in succession, and which depend only on the degree of heat to which they are exposed, that is, on the quantity of caloric which penetrates them; that (2) it is very likely that air is naturally a vaporous fluid, or better put, that our atmosphere is a composition of all fluids that can exist in a vaporous state of constant elasticity, at what is, for us, ordinary heat and pressure; and that (3) it would consequently not be impossible to come across otherwise extremely dense substances in our atmosphere, even metals, such as, for example, a metallic substance just a bit more volatile than mercury.

2.7   Among known fluids, some, such as water and alcohol (i.e., spirit of wine), can mix with each other in any proportion. Others, on the contrary, such as mercury, water and oil, can only combine with one another momentarily, and they separate when mixed and arrange themselves according to their specific gravities. The same thing must, or at least can, occur in the atmosphere.

2.6 *it would ... not be impossible to come across otherwise extremely dense substances in our atmosphere*: The density of a substance is measured by the ratio of its weight to the volume it occupies. All the constituents of the atmosphere are in the gaseous state, and must therefore have very low densities in comparison to solids and liquids. But it is possible that some of those gaseous components, if they could be condensed into liquid form, would prove to be metals like mercury, which is more than 13 times as dense as water. Lavoisier will examine the constitution of atmospheric air in the next chapter.

a metallic substance just a bit more volatile than mercury: Mercury exists naturally in the liquid state; but if some metal were only slightly more volatile than mercury, the atmosphere would be likely to contain significant quantities of that metal's vapor.

2.7 *arrange themselves according to their specific gravities*: The specific gravity of a substance is the ratio of its density to the density of water. In an assortment of fluids that do not mix—or even a solid placed in a fluid—the denser material will displace the less dense as it descends. Thus water, with a specific gravity of 1, forms a layer upon the surface of mercury, whose specific gravity is 13.5; Lavoisier observed this at CD in Plate VII, figure 5 (page 22 above). For the same reason olive oil, with a specific gravity of .93, will form a layer upon the surface of water. But such layers would not form if the liquids could mix with one another.

It is possible, even probable, that gases that mix with atmospheric air only with difficulty and tend to separate from it were formed in the very beginning and form constantly since. If these gases are lighter than air, they must collect in higher regions and form layers which float on atmospheric air. The phenomena associated with fiery meteors lead me to believe that there is a layer of flammable fluid like this above the atmosphere, and that the aurora borealis and other fiery atmospheric phenomena take place at the point of contact between these two layers of air. I intend to develop my ideas on this matter in a separate memoir.

2.7, cont. *gases that mix with atmospheric air only with difficulty*: If the particles of gases are as distant from one another as Lavoisier has proposed (1.44), it is hard to understand what sort of "difficulty" could hinder any gas from mixing with any other gas. Nevertheless, such phenomena are observed. Propane gas, widely used for cooking and heating, is denser than atmospheric air under identical conditions. While propane does mix with air, such mixing takes time, especially indoors where there is no wind. To whatever extent the gas has not diffused, it tends to collect in basements, presenting an explosion hazard. Natural gas (largely methane) is less dense than air and so tends to be displaced upwards and diffuse more readily.

aurora borealis: Northern lights. Their underlying causes remain mysterious even today, but they certainly involve some sort of electrical excitation rather than the fiery combustion Lavoisier suggests.

Chapter Three

The Analysis of Atmospheric Air: Its Resolution into Two Elastic Fluids, One Respirable and the Other Non-respirable

3.1   This much can be said about the constitution of the atmosphere *a priori*: it should be made up of a collection of all gases which remain in an aeriform state at what is, for us, ordinary temperature and pressure. These fluids should form a nearly homogeneous mass from the surface of the earth to the greatest height that has been reached, and the density of this mass should decrease in inverse proportion to the weight it bears. But as I have already said, it is possible that this first layer of the atmosphere is covered with another very different fluid, or several others.

3.2   It now remains for us to determine the number and the nature of the elastic fluids which compose this lowest layer of the atmosphere, the layer we inhabit. Here experiment will light the way. Modern chemistry has made great strides in this context, and the points I will relate will make it clear that atmospheric air is perhaps the most precisely and rigorously analyzed substance of its class.

3.3   In general, chemistry furnishes two means for determining the nature of the constituent parts of bodies: composition and decomposition. When, for example, water and spirit of wine (i.e., alcohol) are combined, creating as a result the liquid known commercially as eau-de-vie, it should be inferred that

---

3.1 *the density ... should decrease in inverse proportion to the weight it bears*: Expressed more straightforwardly, the density of this collection of gases should *increase* in *direct* proportion to the weight that bears on it.

Lavoisier is drawing on the law of compressibility of gases that had been discovered by Boyle in England, and in France by Marriotte: *when the temperature of a gas is kept constant, the volume of the gas is inversely proportional to the pressure to which it is subjected*. Then since the density of a given weight of gas is inversely as its volume (2.6, comment), and the pressure is likewise inversely as the volume, the density and the pressure should be proportional to one another.

Lavoisier is also aware of a relation between a gas's volume and its *temperature*, a relation that would later be known as Charles' Law. He will employ that relation in the measurements mentioned in paragraph 3.7.

eau-de-vie is composed of alcohol and water. But the same conclusion can be reached through decomposition, and in chemistry, complete satisfaction requires, in general, the combination of both of these types of proof.

3.4   In the analysis of atmospheric air, we have the advantage of being able both to decompose and to recombine it. I will limit myself here to reporting the most conclusive experiments that have been made in this regard. Almost all of them are my own, either because I was the first to make them, or because I repeated them with the new aim of analyzing atmospheric air.

3.5   I took a flask, A (Plate II, Figure 14), about 36 cubic inches in volume, with a very long neck, BCDE, about 6 or 7 lines in interior diameter. I bent it in the manner illustrated (Plate IV, Figure 2) so that when placed in the

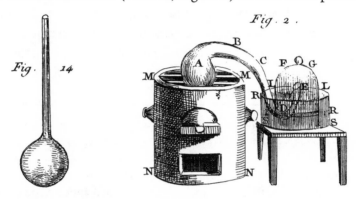

furnace, MMNN, the top of its neck, E, could remain inserted beneath a bell jar, FG, which had been placed in a bath of mercury, RRSS. I introduced into this flask four ounces of very pure mercury, and then, by sucking on a siphon that I had inserted beneath the bell jar, FG, I raised the mercury to the level LL. I glued a band of paper to the bell jar in order to mark this height carefully and took precise observations of the barometer and thermometer.

3.6   With this equipment in place, I lit a fire in the furnace MMNN, and I tended it almost continuously for 12 days, keeping it just below the boiling temperature of mercury.

3.7   Nothing remarkable occurred on the first day. The mercury, though it did not boil, constantly evaporated. It covered the interior of the vessels with droplets, which were at first very fine, but which gradually enlarged. Once they had acquired a certain volume, they fell back again and rejoined the rest of the mercury at the bottom of the vessel. On the second day, I began to see little red particles swimming on the surface of the mercury, which increased in number and volume for about four or five days. After this, they

---

3.5 *the level LL*: In Figure 2, the letters LL actually mark the rim of the mercury *container*. The mercury level is below LL and also below end E of the flask's neck.

stopped increasing and remained in exactly the same state. At the end of 12 days, once I saw that the mercury was no longer calcinating, I extinguished the fire and allowed the vessels to cool. The volume of air contained in the neck of the flask and under the empty part of the bell jar was about 50 cubic inches before the procedure, corrected for a pressure of 28 inches and a temperature of 10 degrees. After the procedure, the same air at equal pressure and temperature was found to be no more than 42 or 43 cubic inches. And so there was about a $\frac{1}{6}$ decrease in the volume of air. Furthermore, after I carefully gathered the red particles that had formed and removed as much of the liquid mercury in which they had been floating as I could, I found their weight to be 45 grains.

3.8   I had to repeat this calcination of mercury many times in closed vessels, because it is difficult, in the same experiment, to preserve both the air in which the experiment is performed and the red particles (i.e., the particles of mercury calx) that are formed. And in what follows I will often assimilate,

---

3.7 *the mercury was no longer calcinating*: As a transitive verb, calcination means the operation of *roasting*. But here the verb is intransitive and denotes the process that occurs when a metal is thoroughly roasted; Lavoisier will relate the details of that process in paragraph 7.2. The substance produced by calcination of any material was commonly called a *calx*. Thus the "little red particles" that appeared on the mercury surface represent the *calx of mercury*, and he will so refer to them in paragraph 3.8. When the particles of calx stop forming, he infers that calcination has ceased. But in the case of metals, Lavoisier will eventually ban the term "calx" in favor of "oxide" (7.4).

corrected for a pressure of 28 inches and a temperature of 10 degrees: A given quantity of substance in the gaseous state does not occupy a fixed volume; rather its volume depends both on the *pressure* to which it is subjected and on its *temperature* (3.1 and comment). Comparison of gas volumes is only meaningful, therefore, if those volumes reflect identical conditions of temperature and pressure.

But gases cannot always be measured under identical conditions. Lavoisier therefore uses the gas relations to calculate what his measured volumes would be had the measurements been performed at 28 inches pressure and 10 degrees (Réaumur) temperature; he calls those calculated results the "corrected" volumes.

45 grains: The pre-revolutionary grain was equal to about .053 grams; so the quantity of calx formed in this experiment was about 2.39 grams.

3.8 *I will often assimilate ... the results of two or three experiments*: In the present instance, for example, he reported that the air volume decreased from an initial 50 to a final 42 or 43 cubic inches as 45 grains of red calx were formed. But he probably had to carry out the experiment twice: once to retrieve the calx (but letting the air escape), and again to collect the air (but perhaps destroying some of the calx).

in this way, the results of two or three experiments of the same kind in the same account.

3.9   The air which remained after this procedure, and which had been reduced to $\frac{5}{6}$ of its initial volume by the calcination of mercury, was no longer fit for breathing or combustion. Animals placed in it died in a few minutes, and flames were immediately extinguished in it as if they had been immersed in water.

3.10   Furthermore, I took the 45 grains of red matter that formed during this procedure and put them in a very small glass retort to which was attached a device for receiving anything liquid or aeriform that might separate out. I lit a fire in the furnace, and I observed that as the red material was heated its color intensified. And when the retort approached incandescence, the red matter began to lose volume bit by bit, disappearing entirely in a few minutes. At the same time, about $41\frac{1}{2}$ grains of liquid mercury condensed in the little receiver, and 7 or 8 cubic inches of elastic fluid passed under the bell jar and was much more fit for combustion and animal respiration than atmospheric air.

3.11   I then collected a portion of this air by passing it into a glass tube with a one inch diameter and immersed a candle in it. The candle gave off a dazzling light. Charcoal, instead of burning quietly as it would in ordinary air, burned like phosphorus in this air, with a flame and a kind of crackling, and with

---

3.10 *a device for receiving anything liquid or aeriform*: Lavoisier's descriptive phrase here lacks the completeness he usually offers. What is this "device?"

the little receiver: Unlike the vessel called by the same name in (1.16), this one is "receiver" in the original sense, a collector of the products of distillation. There is no explicit representation of it in the *Treatise*, but it resembles a piece of glassware drawn in Plate IV, figure 1, upon which this sketch is based. The red calx was placed in the "very small glass retort" A, joined at B to the "little 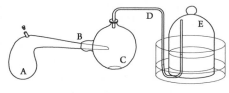 receiver" C. The tube D leads to a bell jar, E, immersed in a water trough.

Initially the bell jar was filled nearly to the top with water. The heated calx gradually disappeared, while liquid mercury collected in the recipient C. At the same time, a gaseous fluid displaced the water in E, lowering its level.

3.11 *a portion of this air*: That is, a portion of the elastic fluid that accumulated in E.

Charcoal ... burned like phosphorus: Phosphorus, an ingredient in matches, burns furiously in air. Pure phosphorus can even ignite spontaneously and is then extremely difficult to extinguish.

an intensity difficult for the eyes to bear. Priestley, Scheele and I discovered this air almost simultaneously. Priestley called it "dephlogisticated air" and Scheele called it "empyreal air." I first gave it the name "highly respirable air," but later this was changed to "vital air." We shall soon see what should be thought of these names.

3.12   On further consideration of the details of this experiment, it is clear that mercury, when it calcinates, absorbs the wholesome and respirable part of the air, or, to speak more rigorously, the base of this respirable part, and that the remaining portion of air is a kind of noxious gas, a gas which cannot

---

dephlogisticated, empyreal, vital: It is interesting to compare these names. Priestley's is a theoretical name, expressing what he took to be the absence of the supposed principle of combustion, *phlogiston*. In contrast, Scheele's and Lavoisier's adjectives express properties that can be directly observed with few, if any, extraneous suppositions. "Empyreal" (French *empiréal*) is Lavoisier's rendering of Scheele's term, "fire" (German *feuer*), and derives from the Greek ἔμπυρος, fiery; this new kind of air supports *fire* of surpassing intensity. Lavoisier's adjectives "respirable" and "vital" express the new substance's ability to maintain or support life. But the paragraph's final sentence hints at Lavoisier's dissatisfaction with *all* these characterizations, his own included.

3.12 *mercury, when it calcinates, absorbs...*: How does Lavoisier know that it *absorbed* something? Let us review the developments:

(i) the air in the jar diminished by 7–8 cu. in., leaving 42–43 cu. in. of a noxious gas residue. Presumably the 7–8 inches of lost air must have gone *somewhere*.

(ii) 45 grains of red calx formed on the mercury surface. But does that calx represent mercury to which *something has been added*, or mercury from which *something has been removed*? Even if we have reason to think the former, it is still only a guess that the *substance added to mercury*, and the *7–8 cu. in. of lost air*, are the same.

(iii) But the calx, when heated intensely, was evidently converted to 41.5 grains of liquid mercury and 7–8 cu. in. of a gas that strongly supports combustion and respiration. It therefore seems certain that the calx represents a *combination* of those two ingredients; and it cannot be mere coincidence that the 7–8 cu. in. of gas lost in the first step deprived the remaining air of respirability, while the 7–8 cu. in. of gas produced at the end exhibited exceptional respirability. Therefore "mercury, when it calcinates, absorbs the wholesome and respirable part of the air."

or, to speak more rigorously, the base of this respirable part: Recall that the gaseous state of a substance consists of (i) the substance itself, which constitutes the *base* of the gas (1.28), and (ii) sufficient caloric to maintain the particles of the base substance far enough apart from one another to erase their mutual attraction.

noxious: Harmful, poisonous, injurious, or generally unwholesome; from Latin *noxa* (injury or harm).

support combustion and respiration. Thus, atmospheric air is composed of two elastic fluids of different and virtually opposite natures.

3.13   One proof of this important truth is that when these two elastic fluids that have been separated from one another (the 42 cubic inches of the noxious gas or non-respirable air, and the 8 cubic inches of respirable air) are recombined, the air that is created is in every way like atmospheric air. It supports combustion, the calcination of metals, and the respiration of animals to almost the same degree.

3.14   Although this experiment provides an extremely simple method for obtaining separately the two principal elastic fluids that constitute our atmosphere, it does not give us precise ideas of the proportion of these two fluids. The affinity that mercury has for the respirable part of the air, or rather for its base, is not strong enough to completely overcome the obstacles that impede this combination. These obstacles are the adherence of these two constitutive fluids of atmospheric air to one another and the force of affinity that unites the base of vital air to caloric. Consequently, when the calcination of mercury is complete, or when, at least, it can proceed no further in a particular quantity of air, a little respirable air remains combined with noxious gas and the mercury cannot separate this last portion. I will later show that the

---

3.13 *when these two elastic fluids ... are recombined*: When he mixes the 42 cu. in. of noxious air with the 8 cu. in. of highly respirable air, the mixture has the same properties as ordinary atmospheric air.

3.14 *it does not give us precise ideas of the proportion*: It gives us an estimate, namely 8 parts respirable to 42 parts non-respirable air (a percentage of $8 \div 50$ or 16%). But, he promises, he will presently deduce a more accurate ratio, 27:73 (equivalent to $27 \div 100$ or 27%). Neither ratio agrees with the value presently accepted, which is 21%. But surely the non-quantitative fruits of this exercise far outweigh its quantitative deficiency! That the air we regard as our essential medium of life—upon which we depend more urgently than on food—is not life-supporting *through-and-through* but rather owes its character to a *specifically vital ingredient* is a discovery that might, in a small way, be compared to our realization that the earth is not the center of the universe.

the affinity that mercury has for the respirable part of the air ... is not strong enough: In order for mercury to capture the respirable portion of the air, it must overcome the alliance of the two parts of air with one another, and also outweigh the bond between the "base" of respirable air and the caloric that endows it with the gaseous state. If its affinity for (attraction to) the base of the respirable portion is weak, it will not fully extract the respirable part. Only at a later stage will Lavoisier be able to explain how he determined that mercury's affinity for the respirable portion of air is, in fact, too weak for that task.

ratio of respirable air to non-respirable air in atmospheric air is 27 to 73, at least in the climates we inhabit. And when I do, I will also discuss the reasons why it remains uncertain whether this ratio is precise.

3.15   Since air is decomposed when mercury calcinates, and the base of the respirable part of the air combines with the mercury, it follows from the principles I have previously outlined that caloric and light ought to be released. And there can be no doubt that they are. But there are two reasons that the light and caloric released will be imperceptible in the experiment I just described. First of all, the calcination lasts several days, which means that light and caloric are released over such a long interval of time that it is extremely weak at any particular moment. Second of all, because the procedure is performed with fire in a furnace, the heat generated by calcination and the heat of the furnace merge. I could add that when the respirable part of the air, or rather its base, forms a new compound with mercury, it does not relinquish all the caloric combined with it, part of it remaining bound to the new compound. But this discussion, and the proofs it would require me to relate, would be out of place here.

3.16   It is, moreover, easy to make the release of heat and light perceptible, by carrying out the decomposition of air more quickly. Iron, which has much more affinity than mercury with the base of the respirable part of air, provides the means for this. By now, everyone knows the beautiful experiment of Ingenhousz on the combustion of iron. Take the end of a very fine iron thread, BC, twisted into a spiral, and attach the end B into a cork stopper, A, made for a bottle DEFG (Plate IV, Figure 17). To the other end of the iron wire, attach a small bit of tinder, C. With all of this in place, fill DEFG with air from which the non-respirable part has been removed. Light the tinder, and

---

3.14, cont. *at least in the climates we inhabit*: Lavoisier evidently suspects that the ratio between the respirable and the non-respirable portions of the air, whatever it may be, is not a *fixed* ratio but may depend, for example, on climate or elevation.

I will also discuss the reasons why it remains uncertain: In paragraph 3.24.

3.15 *when mercury calcinates ... caloric and light ought to be released*: When the base of respirable air combines with the mercury to form red calx, it leaves the gaseous state and must therefore expel caloric. And in paragraph 1.9 Lavoisier noted that caloric and light regularly accompany one another.

3.16 *Iron ... has much more affinity than mercury with ... the respirable part of air*: Iron shows this great affinity by its vigorous combustion in Ingenhouz's experiment, shortly to be described.

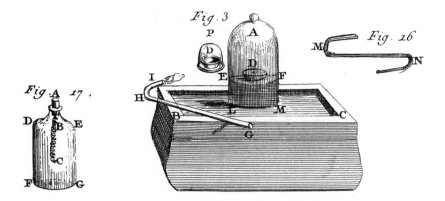

then promptly plug the bottle so that the iron thread BC is inside of it as in the figure I have already cited.

3.17   As soon as the tinder is placed in the vital air, it bursts into a dazzling flame, it communicates the fire to the iron, which also burns, sending out brilliant sparks which fall to the bottom of the bottle in round droplets which turn black as they cool, and which retain a residual metallic luster. The iron burned in this way is even more brittle and fragile than glass. It is easily reduced to a powder, and magnets still attract it, though less so than before combustion.

3.18   Ingenhousz did not examine what happened to the iron or to the air in this procedure, so I found myself obliged to repeat the experiment under different conditions and with equipment better suited to my purposes.

3.19   I filled a bell jar, A (Plate IV, Figure 3), with about 6 pints of pure air, that is to say, with the "highly respirable" part of the air. I carried this bell

---

3.17 *the iron ... burns, sending out brilliant sparks*: Such vehement combustion indicates a strong affinity of iron for the base of vital air. If mercury cannot similarly be set aflame in purely vital air, one would conclude that its affinity for the respirable part of air is weaker (3.16).

3.19 *I filled a bell jar ... with about 6 pints of pure air*: An unfilled jar already contains ordinary air, so how can he fill it with pure air? He first fills the jar with water by immersing it in a tub of water. Then, holding the jar mouth down as shown in the drawing, he uses a suitably bent tube to bubble pure air into the jar; the pure air replaces the water in the jar.

pure ... "highly respirable": Lavoisier is still casting about for a suitable name. He will propose one in the following chapter, after having discussed the principles that will govern his naming of newly-discovered substances.

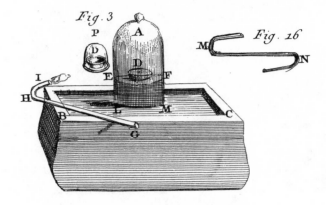

jar over to a bath of mercury contained in a basin, BC, with the aid of a very shallow bowl, after which I carefully dried the surface of the mercury with scraps of blotting paper, both inside and outside the bell jar. Furthermore, for the purposes of this experiment, I used a little porcelain saucer, both flat and flared at the lip, in which I had placed iron shavings twisted into spirals and arranged in a manner which seemed to me most likely to allow the combustion to spread to all of the chips. I attached a bit of tinder to the end of one iron chip, and to this I added a fragment of phosphorus which hardly

---

3.19, cont. *I carried this bell jar over to a bath of mercury ... with the aid of a very shallow bowl*: While the mouth of the jar (now filled with pure air) is still under water, he slides the bowl beneath it, as in the drawing. When he lifts both jar and bowl out of the tub, enough water remains in the bowl to seal the opening and thus retain the pure air in the jar. He carries both jar and bowl to a bath of mercury, immerses them in the mercury bath, and removes the bowl. His Figure 3 depicts the result.

I carefully dried the surface of the mercury: Since water is less dense than mercury, the water brought over in the shallow bowl inevitably floats upon the mercury surface, both inside and outside the jar. He uses blotting paper, which soaks up water but not mercury, to dry it. How might he manipulate the blotting paper so as to dry the mercury *within* the jar?

Why could he not have filled the bell jar over mercury at the outset, and so avoid bringing water into the mercury apparatus at all? The tub into which the bell jar was immersed would then have had to contain mercury, which is nearly fourteen times heavier than the same quantity of water. Lavoisier indicates in Part III of the complete Treatise that such weight was more than any wooden tub could support.

I added a fragment of phosphorus: Phosphorus is strongly disposed to ignite even in common air, much more so in highly respirable air. Along with the "bit of tinder," it serves as a fire-starter to help kindle the iron shavings.

weighed $\frac{1}{16}$ of a grain. I placed the saucer under the bell jar by lifting the end a little. I realize that when the experiment is performed this way, a little bit of common air mixes with the air under the bell jar, but when one works quickly, this mixing in no way compromises the experiment's success.

3.20   Once I had placed the saucer D under the bell jar, I sucked out part of the air from the bell jar in order to raise the mercury inside it to the height EF. A siphon, GHI, should be used for this purpose. It should be plugged with a wad of paper to keep it from filling with mercury, and then inserted into the bell jar from below. There is an art to sucking air in such a way as to raise the mercury under a bell jar. If you just use your lungs, you can only raise it a little, perhaps an inch, an inch and a half at most. But if you also use the muscles of your mouth, it is possible to raise the mercury about 6 or 7 inches without tiring yourself, or at least without too much inconvenience.

3.21   After all of this was prepared, a bent piece of iron, MN, (Figure 16) designed for experiments like these, is heated until it glows red. It is then inserted into the bell jar from below; and before it has had time to cool, it is brought close to the bit of phosphorus contained in the porcelain saucer, D. As soon as the phosphorus catches fire, it communicates its flame to the tinder, which in turn communicates it to the iron. When the iron chips have been arranged well, a brilliant white flame, like Chinese fireworks, spreads

---

3.20 *Once I had placed the saucer under the bell jar*: Saucer D, containing the iron shavings, *floats* on the mercury surface.

It should be plugged with a wad of paper to keep it from filling with mercury: The plugged end I of the siphon is passed under the jar, through the mercury, and into the highly respirable air above the mercury surface. Remarks that Lavoisier will make in paragraph 3.24 indicate that the plug is not airtight; he closes end G of the tube with a finger while the plugged end is immersed.

use the muscles of your mouth: In using the mouth rather than the lungs to create suction, Lavoisier would have reduced only slightly the danger of inhaling toxic mercury vapor. Such considerations were surely no part of his thinking; but the technique is a risky one. Today, anyone working with mercury would use a mechanical aspirator to perform that task.

3.21 *before it has had time to cool*: One end of the iron wire is grasped with a suitable tool; the other end is plunged into the mercury bath, then maneuvered under the jar edge and raised so as to contact the phosphorus. If this is done quickly, the iron will still be hot enough to ignite the phosphorus.

out over the iron and burns every last bit. The great heat produced during this combustion liquefies the iron, which falls in round drops of different sizes. These drops stay in the saucer for the most part, but some jump out and float on the surface of the mercury.

3.22   During the first moment of combustion, there is a slight increase in the volume of the air because of the expansion caused by the heat. But rapid contraction soon follows this expansion. The mercury rises again in the bell jar, and when there is enough iron and the air used is very pure, almost all of the air is absorbed.

3.23   So I should warn readers here that unless one is making experiments for the purposes of research, it is better to burn only modest quantities of iron. For when the experiment is pushed too far and almost all the air is absorbed, the saucer D which floats on the mercury comes too close to the top of the bell jar, and the great heat together with the sudden cooling caused by contact with the mercury, shatters the glass. And as soon as there is a crack in the bell jar the weight of the column of mercury causes it to fall very quickly, and this leads to such a flood, that most of the mercury gushes out of the basin. In order to avoid these inconveniences, and to be sure that the experiment succeeds, no more than 1.5 gros of iron should be burned under a bell jar with a volume of 8 pints. And this bell jar should be strong enough to hold up against the weight of the mercury it will contain.

3.24   In this experiment, it is impossible to determine the weight the iron acquires and the changes that occur in the air at the same time. If the aim is to determine the weight of the iron and its relationship to the absorption of air, be sure to mark the height of the mercury on the bell jar very exactly with the tip of a diamond both before and after the experiment. Then pass the siphon, GHI, which has been plugged with paper to prevent it from filling with mercury, under the bell jar. Place a thumb on one end, G, and then allow air in under the thumb a little at a time. When the mercury has

---

3.22 *when ... the air used is very pure*: That is, when it contains the respirable part of air almost exclusively.

    almost all of the air is absorbed: The same reasoning that implied "absorption" of vital air by mercury (3.12) will apply to the iron shavings used here; and, evidently, a very small quantity of iron is capable of "absorbing" nearly 6 pints (3.19) of vital air.

3.23 *no more than 1.5 gros of iron should be burned*: 1.5 gros is equivalent to 5.7 grams. For this and other equivalents, see the table of pre-revolutionary weights and measures in the Appendix.

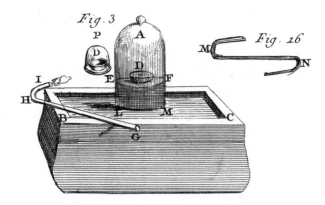

descended to its level, slowly lift the bell jar, remove the droplets of iron from the saucer, and carefully gather any that have splashed out and are floating on the surface of the mercury. Then weigh all the iron. This iron is in a state that earlier chemists called "martial ethiops." It has a kind of metallic luster, is very brittle, crumbles easily, and can be reduced to a powder with mortar and pestle. When the procedure works well, 135 or 136 grains of martial ethiops can be obtained from 100 grains of iron. So the increase in weight is at least 35 parts per hundred.

3.25   If this experiment has been carried out with the careful attention it deserves, the weight of the air diminishes by an amount exactly equal to the increase in the weight of the iron. So if 100 grains of iron are burned and the iron gains 35 grains, the volume of air will diminish, at a rate of half a grain per cubic inch, by almost exactly 70 cubic inches. We will see in later chapters

---

3.25 *the weight of the air diminishes by an amount exactly equal to the increase in the weight of iron:* Therefore the total quantity of matter (evidenced by its weight) involved in the combustion is *constant*. This is a general principle for Lavoisier, which, remarkably, he does not state explicitly until Chapter Thirteen. There he writes: "For nothing can be created by either natural or artificial processes. The amount of matter at the outset of any process is equal to the amount of matter that remains afterwards..." Lavoisier was by no means the first to propound the principle of *conservation of mass*, as it is called today. But he was the first to treat it as not just a philosophical truism but a practical tool of analysis. Scrupulous determination of weights, using a superbly-crafted balance, made this possible. The photo shows the analytical balance made for Lavoisier by Fortin in 1788, in the collection of the Musée des Arts et Métiers, Paris. Used by permission.

of this memoir that the weight of vital air is in fact almost exactly half a grain per cubic inch.

3.26   I will remind readers here for the last time, that in all experiments like this one, it is necessary to recalculate the volume of air, both at the beginning and at the end of the experiment, in order to determine what it would have been at a temperature of 10 degrees and a pressure of 28 inches. I will explain in some detail how to make the necessary corrections at the end of this work.

3.27   If the aim of the experiment is rather to investigate the quality of the air that remains in the bell jar, a somewhat different approach is required. After the combustion is complete and the vessels have cooled down, begin by removing the saucer that contains the iron by passing your hands under the bell jar and through the mercury.... Finally, introduce enough water into the bell jar to displace all the mercury, and pass a vessel such as a very flat saucer under the bell jar so that it can be carried over to a common pneumato-chemical apparatus for water, with which it is possible to work on a larger scale and with greater facility.

3.28   When very soft, very pure iron is used, and when the combustion takes place in respirable air without any mixture of non-respirable air, the air that remains after combustion will be found to be as pure as it was before combustion. But it is rare for the iron not to contain a small quantity of charcoal matter. Steel always contains some. It is also extremely difficult to obtain perfectly pure respirable air. This air is almost always mixed with a little bit of non-respirable air, but this kind of noxious gas in no way compromises the results of the experiment, and at the end of the experiment the same amount of it that was there in the beginning remains.

---

half a grain per cubic inch: This is Lavoisier's figure for the density of highly respirable air at the standard (for him) pressure and temperature of 28 inches of mercury and 10° Réaumur. Lavoisier reserves the description of his methods for determining weights and volumes of gases for Part III of the complete Treatise.

3.26 *how to make the necessary corrections*: Lavoisier's explanation is not part of the present selections; his procedure is essentially equivalent to the modern application of Boyle's (or Marriotte's) and Charles' laws (3.1, comment).

3.27 *pneumato-chemical apparatus*: This is the unwieldy name French chemists gave to the *pneumatic trough*, which Joseph Priestley had devised for the collection of gases in a bell jar or other container. Lavoisier's Plate IV, figure 3 (on the previous page) depicts a small example, suitable for collecting gases over mercury. Lavoisier also employed a larger version for use with water; such was the tub referred to in the comment on paragraph 3.19.

3.29   Earlier, I indicated that there are two ways to determine the nature of the constituents of atmospheric air, decomposition and composition. The calcination of mercury furnishes us with an example of both, since, after having removed the base from the respirable part of the air with mercury, we replaced it so as to form air in every way like atmospheric air again. But it is equally possible to affect the composition of atmospheric air by gathering the required materials from different kingdoms of nature. We will see below that dissolving animal matter in nitrous acid releases a great quantity of the kind of air that extinguishes flames, air that is harmful to animals, and which is in every way like the non-respirable part of atmospheric air. If 27 parts of highly respirable air, which has been extracted from red calx (i.e., calcinated mercury), are added to 73 parts of this non-respirable air, an elastic fluid exactly like the atmosphere and with all the same properties is formed.

3.30   There are many other ways to separate the respirable from the non-respirable part of the air, but I cannot explain them without appealing to concepts which, in the order of knowledge, belong to later chapters. The experiments I have already reported suffice for an elementary treatise, and in these sorts of matters, it is more important to choose proofs carefully than to present a great number of them.

3.31   I will conclude this article by mentioning a property of the atmosphere which, in general, all elastic fluids or gases that we know of share: that of dissolving water. A cubic foot of atmospheric air can dissolve about 12 grains of water, according to the experiments of Saussure. Other elastic fluids, such as carbonic acid, can dissolve more, but the experiments to determine exactly

---

3.29 *nitrous acid*: Lavoisier will discuss, and name, this acid in Chapter Six.

exactly like the atmosphere: Since (i) the combination has the same properties as common air, and (ii) one of the ingredients is known at the outset to be the respirable part of common air, then (iii) the second ingredient would appear to be identical to the non-respirable part of the air—even though it was obtained from a source that has no obvious connection to the atmosphere.

3.31 *dissolving water*: The most common cases of "dissolving" are those that feature a solid or liquid *solute* (sugar or milk) and a liquid *solvent* (coffee or tea). But Lavoisier does not hesitate to extend ordinary usage to include the case of a liquid solute (water) and a gaseous solvent (air). And he has already compared caloric—the paradigmatic elastic fluid—to a solvent (1.28).

carbonic acid: This acid will be discussed in Chapter Five and named in Chapter Six.

how much have not yet been made. The water that elastic, aeriform fluids contain gives rise to certain phenomena in some experiments which merit more attention and which have often thrust chemists into great errors.

---

great errors: One such error can arise when gases are collected over water in the pneumatic trough. The desired gas will inevitably harbor some water vapor, which in some cases will lead to error if not taken into account. But no such complications will arise in the experiments that appear in the present selections.

Chapter Four

A Nomenclature for the Different Constituent Parts of Atmospheric Air

4.1   So far, I have been forced to use periphrases to designate the nature of the different substances which compose our atmosphere. I have provisionally adopted expressions such as "the respirable part of the air" and "the non-respirable part of the air." But I am about to go into much greater detail, and this calls for a more rapid pace. I have already tried to convey simple ideas of the different substances which go into the composition of atmospheric air. From now on, I will use words equally simple to refer to them.

4.2   The temperature of the planet we inhabit is found to be very close to the degree at which water passes back and forth between a liquid and a solid state. This phenomenon frequently occurs before our very eyes, and so it is not surprising that in all languages, at least those spoken in climates where they have winter, there is a name for water that has become solid through lack of caloric.

4.3   But the same cannot be said for water converted to a vaporous state through the absorption of a great quantity of caloric. Those who have not made a special study of these matters still do not know that, at a temperature slightly higher than that at which it boils, water transforms into an elastic, aeriform fluid which, like all gases, can be captured and contained in vessels, and which keeps its gaseous form so long as it is exposed to temperatures greater than 80 degrees under a pressure equal to that of a 28 inch column of mercury. Since this phenomenon has escaped the notice of most people, no language has a word for water in this state, and it is the same for all fluids and, in general, for all substances which never evaporate at what are for us ordinary temperatures and pressures.

4.2 *there is a name for water that has become solid through lack of caloric*: In English that name is, of course, *ice*.

4.3 *no language has a word for water in this state*: One might think that both the French "*vapeur*" and English "steam" are such words. But the "state" Lavoisier is speaking of is a pure, invisible gas; whereas, in popular use, both these terms refer to substances like the exhalations of kettles, which form visible clouds because they contain minute droplets of liquid water. Since the eighteenth century, however, both terms have acquired Lavoisier's meaning in scientific and technical contexts.

4.4   For the same reason, there is no word for most aeriform fluids in the liquid or solid state. For no one knew that these gases were the result of the combination of a base with caloric, and since they had never been seen in either a liquid or a solid state, no one, not even natural philosophers, knew that they existed in these forms.

4.5   We decided that it was not open to us to change conventional names which had been thoroughly established in society through long use. For this reason we have retained the colloquial senses of the words "water" and "ice." We have even used the word "air" to signify the collection of elastic fluids that compose our atmosphere. But we did not consider ourselves similarly obliged to adhere to the very modern names recently proposed by natural philosophers. We thought instead that we were right to reject these names and replace them with others less likely to lead to errors. Even when we decided to adopt recently proposed names, we had no qualms about modifying them or attaching them to more distinct and restricted ideas.

4.6   We have drawn these new words principally from the Greek, and we have chosen them so that their etymologies call to mind the ideas of the things we intend to signify. Above all else, we were committed to using short words and, whenever possible, short words from which adjectives and verbs could be formed.

4.7   In accordance with these principles, we have followed the example of Macquer and have employed the name "gas" used by van Helmont; and we have subsumed the large class of elastic, aeriform fluids under this name, with the exception of atmospheric air. Thus, the word "gas" as we use it is a generic name which designates the last degree of saturation of some substance by caloric; it is a word that expresses a mode of corporeal being. It was then necessary to find a specific name for each kind of gas, and we did this by deriving a second word from the name for the base of each gas. Thus, we call water combined with caloric in the state of an elastic, aeriform fluid "aqueous gas." We call ether combined with caloric "ethereal gas," spirit of wine combined with caloric, "alcoholic gas," and similarly for "muriatic acid gas," "ammoniac gas," and all the others. I will further expand this list of names for different gases when I provide the names for their different bases.

4.4 *combination of a base with caloric*: Recall Lavoisier's usage of "base," discussed in the comment to paragraph 1.28.

4.7 *a mode of corporal being*: That is, a *state* or *condition* of body, rather than a fixed property.

4.8   We have seen that atmospheric air is principally composed of two aeri-form fluids or gases. One is respirable and capable of sustaining the life of animals. Metals calcinate in it, and combustible bodies burn. The other has absolutely opposite properties: animals cannot breathe in it, it cannot sustain combustion, and so on. We named the base of the respirable part of the air "oxygen," which we derived from two Greek words: ὀξύς, ("acid") and γείνομαι ("I generate") because one of the most common properties of this base is to form acids when it combines with most substances. So when this base unites with caloric we call the result "oxygen gas." Its weight in this state is almost exactly half a standard grain per cubic inch or an ounce and a half per cubic foot at a temperature of 10 degrees and a barometric pressure of 28 inches.

4.9   Since the chemical properties of the non-respirable part of atmospheric air are still not well known, we were content to derive the name of its base from this gas's property of killing animals that breathe it. We have for this reason called it "azote" using the Greek alpha privative and the word ζωή, "life." And so the non-respirable part of the air will be called "azotic gas." Its weight is 1 ounce, 2 gros, 48 grains per cubic foot or 0.4444 grains per cubic inch.

4.10   We realize that this name seems rather strange. But this is the way that all new names seem, and it is only through use that we become accustomed to them. We spent quite some time exploring alternatives, but it proved

---

4.8 ὀξύς (*acid*): The Greek adjective means *sharp, acute,* or *pungent*. Lavoisier will demonstrate the role of oxygen as "acid-maker" in Chapters Five and Six.

4.9 *Since the chemical properties ... are still not well known...*: Ideally, the name of a simple substance should reflect its chemical properties; when these are unknown, other properties must serve as the basis for naming. But what distinguishes "chemical" properties from other kinds of properties?

...we were content to derive the name [*azote*] *from this gas's property of killing animals that breathe it*: Evidently Lavoisier does not regard azote's *inability to support respiration* as a chemical property, in contrast to oxygen's *ability to form acids*, which he does view as "chemical." Now the ability of one substance to form another (as oxygen does) is a relation between a substance and a substance; but the inability of a azote to sustain life is a relation between a substance and an *entire complex organism*. Perhaps, then, a "chemical" property is one that acts at the substance-to-substance level.

the Greek alpha-privative: Numerous Greek words express the negation of something by prefixing the letter *alpha* to its name. English words like *atypical* (meaning "not typical") and other words derived from Greek carry over this formation.

impossible to find a better name than this one. We were at first tempted to call it "alkaligenic gas" in recognition of Berthollet's experiments which proved that this gas is a constituent of volatile alkali (ammoniac), as will be seen in what follows. But on the other hand, we have as yet no proof whatever that it is one of the constituent principles of other alkalis, and, furthermore, it has been shown to be a constituent of nitrous acid, and so we had equal grounds for calling it "the nitrogenic principle." In the end, we had to reject the names which carried with them some idea that was in accordance with our system, and we thought we avoided the risk of making a mistake by adopting the names "azote" and "azotic gas," since these only express a fact, or rather a property that the gas possesses, namely the property of killing the animals that breathe it.

4.11   If I said anything more about the nomenclature of the different kinds of gases, I would need to appeal to notions reserved for later chapters. So I will not provide names for all the gases here. For now, it suffices that I have explained the method used to name them. The chief merit of the system of classification that we have adopted is that the names of all of the compounds of a particular simple substance follow necessarily from that simple substance's name.

4.10 *alkaligenic*: Such a name would express a genuinely chemical property, the *ability to generate alkalis*; it would, therefore, be preferable to "azote."

volatile alkali (*ammoniac*): This substance, which Lavoisier mentioned earlier in paragraph 1.28, is an example of an alkali that contains azote. But since it has so far proved to be the *only* example, we cannot regard the *ability to generate alkalis* as characteristic of azote.

we had to reject the names which carried ... some idea that was in accordance with our system: Ideas such as *ability to generate alkalis* or *ability to generate a class of acids* would have been "in accordance with our system," since either would have represented a chemical property—a capacity of the substance in question to act with respect to other substances (0.32). But since both of those powers could be confirmed for azote only in single instances and not generally, the names based on them had to be discarded.

"azote" and "azotic gas" ... only express a fact, or rather a property that the gas possesses: The name doesn't express a *chemical* property, which would be preferable. But at least it expresses a *fact*; that is, it asserts only so much as we may observe.

Chapter Five

*On the Decomposition of Oxygen Gas by Sulphur, Phosphorus, and Charcoal,
and on the Formation of Acids in General*

5.1   It is a necessary principle, which must never be neglected in the art of experimentation, that experiments should be simplified as much as possible; that every circumstance that could complicate the results should be eliminated. Thus, in the experiments treated in this chapter, atmospheric air will not be used, since it is by no means a simple substance. It is certainly true that azotic gas, a part of the mixture that constitutes air, appears to be entirely passive in calcinations and combustions. But since it slows down these processes and perhaps even changes their outcomes in certain cases, it seemed to me necessary to eliminate this source of uncertainty.

5.2   Therefore I will only present the results of combustions that took place in vital air (i.e., in pure oxygen gas) in the experiments described below, and I will only note those differences which occur when oxygen gas is mixed with azotic gas in different proportions.

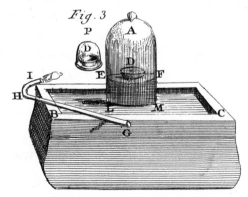

5.3   I took a crystal bell jar, A, with a volume of between five and six pints (Plate IV, Figure 3). I filled it with oxygen gas over water, after which I transported it to a bath of mercury by means of a glass saucer that I placed beneath it. I then dried the surface of the mercury and introduced, under the bell

---

5.3 *I filled it with oxygen gas over water.... I then dried the surface of the mercury*: Lavoisier again uses the filling procedure discussed earlier in the comment to paragraph 3.19. As before, that technique introduces some water into the apparatus and therefore requires him to dry the mercury surface.

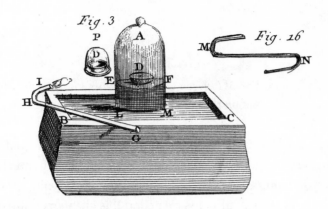

jar A, $61\frac{1}{4}$ grains of Kunkel's phosphorus on two porcelain saucers similar to the one depicted in D, Figure 3. I covered one of the two portions of
phosphorus with a small plate of glass, so that I could light each of the two
portions separately, and so that the flame would not pass from one to the
other. Once all of this had been arranged, I raised the level of mercury in the
jar to height EF by sucking on a glass siphon GHI that had been inserted
beneath the jar. In order to avoid filling the siphon as it passed through the
mercury, I plugged one end with a small wad of paper. Then, with a piece of
curved wire (as depicted in Plate IV, Figure 16) that had been heated in a fire
until it glowed red, I lit the two saucers of phosphorus successively, beginning with the one which was not covered by the square of glass.

5.4   The combustion occurred very rapidly. There was a brilliant flame, and a
considerable amount of heat and light was released. In the first instant, the
oxygen gas expanded considerably, due to the heat. But soon afterwards the
mercury rose back above its level, and a great amount of oxygen was absorbed.
At the same time, the whole interior of the jar was covered with light, white
flakes, which were nothing other than solid phosphoric acid.

5.3 *Kunkel's phosphorus*: Johann Kunkel, though not its discoverer, produced
phosphorus commercially by a proprietary process. The quantity $61\frac{1}{4}$ grains is very
nearly equal to 3.25 grams.

    *with a piece of curved wire ... heated in a fire until it glowed red, I lit the two saucers of
phosphorus*: Lavoisier used this technique earlier in paragraph 3.21.

5.4 *a great amount of oxygen was absorbed*: Since the mercury "rose back above its
level," it is clear that the total amount of oxygen in the jar decreased. By analogy
with his earlier combustion of mercury (3.5) and iron (3.19), it is reasonable to
suppose that the vanished oxygen has been absorbed by the phosphorus. But he
will, as always, confirm this supposition by weighing the materials involved.

    light, white flakes, which were ... solid phosphoric acid: Lavoisier is anticipating here, as

5.5   After all corrections had been made, there were 162 cubic inches of oxygen gas at the beginning of the experiment. At the end, there were only $23\frac{1}{4}$ cubic inches. Thus $138\frac{3}{4}$ cubic inches or 69.375 grains of oxygen gas had been absorbed.

5.6   Not all of the phosphorus burned. Several pieces remained in the saucers. When these were washed so as to separate out the acid, and then dried, they were found to weigh approximately $16\frac{1}{4}$ grains. Thus 45 grains of phosphorus, more or less, were burned. I say "more or less" because it is possible that there was an error of one or two grains with respect to the weight of the phosphorus remaining after combustion.

5.7   Thus in this procedure, 45 grains of phosphorus combined with 69.375 grains of oxygen. And since nothing with weight passed through the glass, we are entitled to conclude that the weight of whatever substance was produced by this combination, and which had collected in white flakes, should be the same as the sum of the weights of the oxygen and the phosphorus, that is,

---

indeed he anticipated earlier (4.8) when he stated that oxygen possesses the ability to generate acids. Not until paragraph 5.17 will he actually show that these white flakes have an acidic character.

5.5 *After all corrections had been made*: That is, after converting the gas volumes measured under actual conditions to the volumes those same quantities of gas would have under Lavoisier's standard conditions of 28 inches pressure and temperature of 10° Réaumur.

138¾ cubic inches or 69.375 grains of oxygen gas had been absorbed: 138¾ is the difference between the corrected initial volume of oxygen and the corrected final volume. And in paragraph 4.8, Lavoisier reported the weight of a cubic inch of oxygen (at his standard pressure and temperature) to be almost exactly one-half grain. Multiply 138.75 by .5 to obtain Lavoisier's figure.

5.6 *these were washed to separate out the acid*: Unburned pieces of phosphorus were coated with some of the white flaky material. As Lavoisier will remark in paragraph 5.17, the white substance is highly soluble in water (indeed, such solubility is one of the typical properties of acids). Phosphorus, on the other hand, does not react with water; so washing is practical. After washing and drying, the unburned phosphorus could be weighed.

45 grains of phosphorus ... were burned: 45 is the difference between the initial weight of phosphorus, 61¼ grains, and the remaining weight, 16¼ grains.

5.7 *nothing with weight passed through the glass*: It is of course certain that *caloric* passed through the glass (5.4); but caloric has no detectable weight.

the weight ... produced ... should be ... the sum of the weights of the oxygen and the phosphorus: Note this implicit use of the principle of constancy of weight, discussed in 3.25, comment.

114.375 grains. We will soon see that the white flakes are nothing other than a solid acid. Expressing these quantities in parts per hundred, we find that 154 livres of oxygen are required to saturate 100 livres of phosphorus, and this produces 254 livres of white flakes, i.e., solid phosphoric acid.

5.8   This experiment proves in a very clear way that oxygen has more affinity for phosphorus than for caloric at a certain degree of temperature. Consequently, phosphorus decomposes oxygen gas, attaches to its base, and then the caloric which is liberated escapes and disperses as it spreads into surrounding bodies.

5.9   But no matter how conclusive this experiment may have been, it was not yet sufficiently rigorous. In fact, it is not possible to verify the weight of the white flakes (i.e., the solid acid) produced in the apparatus that I used and just described. We can only infer this weight by means of a calculation, by assuming that it is equal to the sum of the weights of the oxygen and phosphorus. But however evident this inference may be, in chemistry and physics one must never infer something that can be ascertained through direct experiment.

5.7, cont. *Expressing these quantities in parts per hundred*: 69.375 is to 45 as 154 is to 100.

5.8 *oxygen has more affinity for phosphorus than for caloric…*: Since the oxygen was initially in gaseous form, it must have been united to caloric. But when it combined with the phosphorus it became a solid, thereby giving up considerable caloric. Oxygen must therefore have been attracted to phosphorus more than to caloric.

…at a certain degree of temperature: The phosphorus had to be ignited—that is, brought to a sufficiently high temperature—before oxygen would combine vigorously with it. But phosphorus's tendency to *glow spontaneously* in air suggests that oxygen may have considerable affinity for it even at ordinary temperatures.

phosphorus decomposes oxygen gas: Modern chemistry teaches that oxygen is an elementary substance. And even Lavoisier has no evidence to suggest that it is other than "simple"; how then can it be "decomposed"? Recall that, for Lavoisier, every gas is a compound of that gas's *base* and *caloric*. Thus any gas that becomes either liquid or solid undergoes decomposition.

5.9 *It is not possible to verify the weight of the white flakes*: Notice that Lavoisier made no effort to collect the white flaky material for weighing. Even had he done so, the operation of washing the unburned phosphorus to remove the white coating (5.6) would have caused the loss of an undetermined amount of the white substance.

one must never infer something that can be ascertained through direct experiment: Here, Lavoisier is applying this maxim to his doctrine that *no substance is either created or destroyed* (3.25, comment)—a doctrine that is crucial to his judgement that oxygen was *absorbed* by the phosphorus (5.4). Even though he may have had

I therefore believed it was necessary to perform the experiment again on a somewhat larger scale and with different equipment.

5.10   I took a large spherical glass flask, A, with an opening, EF, three inches in diameter (Plate IV, Figure 4). I covered this opening with a crystal plate deburred with emery paper and into which two holes had been drilled so as to allow the tubes *xxx* and *yyy* to pass.

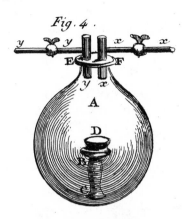

5.11   Before closing up the flask with its plate, I set up a stand, BC, inside of it, on top of which was a porcelain saucer, D, containing 150 grains of phosphorus. When everything had been arranged in this way, I placed the crystal plate over the opening of the flask, and I sealed the opening with a lute made from fat and covered the seal with linen bands soaked in quicklime and egg whites. When the lute was thoroughly dry, I suspended the whole apparatus from the arm of a balance, and I determined its weight within nearly a grain or a grain and a half. Next I fitted tube *xxx* to a small air pump and evacuated the flask. Then I let oxygen gas into the flask by opening a stopcock fitted to tube *yyy*. I will note here that experiments like these can be performed easily enough and,

---

good reason to suppose such absorption by analogy with earlier experiments on combustion, he will repeat the experiment using apparatus that enables him to weigh all the materials involved.

5.10 *a crystal plate deburred with emery paper*: Lavoisier would have cut or sawn the circular crystal (glass) plate EF, whose edges would then require smoothing and sanding. Emery paper is coated with a hard, impure, abrasive mineral, and was used the way "sandpaper" is used today.

5.11 *lute*: A clay or cement for sealing joints; from Latin *lutum*, mud.

quicklime: It becomes a powerful alkali when mixed with water, but here Lavoisier is using it for its sticky, adhesive qualities. Compare Wordsworth: "...a snare for inattentive fancy, like the lime that foolish birds are caught with" (*Sonnet composed upon Westminster Bridge, Sept. 3, 1802*).

I determined its weight: Lavoisier did not do this in the first experiment. Weighing the flask *and its contents* allows him to monitor the total weight of materials involved.

evacuated the flask: He removed from the flask as much common air as possible, in order to replace it with pure oxygen gas.

more importantly, with great precision using the hydro-pneumatic machine Meusnier and I described,* and which is explained in the last part of this work. By means of this instrument, to which Meusnier has since made important additions and improvements, one can ascertain precisely both the quantity of oxygen gas introduced into the flask, and the amount consumed over the course of the procedure.

5.12   Once I had set up the apparatus in the way described above, I set fire to the phosphorus with a burning glass. The combustion was extremely rapid, accompanied by a great flame and a much heat. During combustion, a great quantity of white flakes formed, which attached themselves to the inner walls of the vessel, soon enshrouding it completely. The vapors emitted were so profuse that even though the combustion ought to have been sustained by the new supplies of oxygen gas continually passing into the flask, the phosphorus was soon extinguished. Once I had allowed the whole apparatus to cool off completely, I began by determining the quantity of oxygen gas that

* *Mémoires de l'Académie*, 1782, p. 466.

5.11, cont. *hydro-pneumatic machine*: Also called *gasometer*, this device delivered a measured volume of gas at constant pressure. Since Lavoisier has already determined the weight of one cubic inch of oxygen (3.25), he can calculate the weight of any measured volume. The gasometer is convenient for the present experiment; it will be indispensable for an experiment to be described in Chapter Eight.

5.12 *burning glass*: Lavoisier used a large magnifying glass to focus the sun's rays on the phosphorus, thereby heating it to ignition.

the new supplies of oxygen gas continually passing into the flask: The hydro-pneumatic machine automatically supplied oxygen to the flask in whatever quantity was needed to keep the pressure constant. In so doing, it both replaced and measured the amount of oxygen consumed in combustion.

to cool off completely: He allowed the apparatus to regain, as nearly as possible, the temperature it had at the start of the experiment. The hydro-pneumatic machine remained connected, thereby keeping the pressure constant during this period.

I began...: That is, he began the process of *assessing the results* by carrying out two operations: (i) measuring the amount of oxygen consumed, and (ii) weighing the flask and its contents, including the newly-formed white flakes.

...by determining the quantity of oxygen gas that had been used...: This quantity was measured by the hydro-pneumatic machine (5.11). The machine kept the oxygen pressure constant, and after combustion the oxygen returned to its initial temperature. Then since the conditions before and after combustion were identical, the quantity of oxygen supplied must exactly equal the quantity of oxygen absorbed.

had been used, and by weighing the flask before opening it up. I then washed, dried, and weighed the small quantity of phosphorus which remained in the saucer and which was yellow in color like ochre, so as to deduce the total quantity of phosphorus used in the experiment. It is clear that with the help of these different precautions, it was easy for me to ascertain (1) the weight of phosphorus burned; (2) the weight of the white flakes produced by combustion; and (3) the weight of oxygen gas which had combined with the phosphorus.

5.13   This experiment gave me more or less the same results as the preceding one. Just as before, when the phosphorus burned, it absorbed a little more than one and a half times its weight in oxygen; and in addition I became more certain that the weight of the new substance produced was equal to the sum of the weights of the burned phosphorus and the oxygen it had absorbed, which was, in any case, easy to foresee *a priori*.

...and by weighing the flask before opening it up: He again weighs the flask and its contents, as he did prior to the combustion. Is it obvious what purpose is served by this second weighing?

I then washed, dried and weighed the ... phosphorus which remained: The difference between the initial weight of phosphorus and the weight that remained after combustion is the weight of phosphorus that was burned.

ochre: An earth rich in clay, with color varying from yellow to brown; it is widely used as a pigment.

precautions: On its face, a strange word to use here. A *precaution* is a special measure taken in advance in order to guard against a known risk. But washing and drying the residual phosphorus is merely routine procedure prior to weighing it.

Not so, however, for weighing the flask! Since the total quantity of matter involved in the combustion is constant (3.25), the difference between the final and initial weights of the *flask with its contents* must be equal to the *weight of oxygen supplied*. But Lavoisier made it clear in paragraph 5.11 that he relied on the *hydro-pneumatic machine* to measure the oxygen. Weighing the flask, therefore, appears indeed to be a genuine "precaution," providing an alternative determination of the oxygen consumed; if the two measures fail to agree, that will alert Lavoisier to an error in either his reasoning or his procedure. Compare Lavoisier's phrase "double verification" in (8.28)

it was easy for me to ascertain...: He has both (1) the weight of phosphorus burned, and (3) the weight of oxygen that combined with it, by direct measurement. Then since the total quantity of matter involved in the combustion is constant (3.25), the weight of the white flakes (2) is simply the sum of (1) and (3).

5.13 *a priori*: A logical term, here meaning "in advance."

5.14　　　If the oxygen gas used in this experiment is pure, the residue that remains after combustion is equally pure. This proves that nothing is emitted from the phosphorus that can compromise the purity of the air, and that phosphorus does nothing else but remove the base, which is oxygen, from the caloric to which it is united.

5.15　　　I mentioned above that if a combustible body were burned in a hollow sphere of ice or in any apparatus constructed according to the same principle, the quantity of ice melted during combustion would be an exact measure of the quantity of caloric released.* When we tested the combustion of phosphorus in this way, we discovered that burning 1 livre of phosphorus melted a little more than 100 livres of ice.

5.16　　　The combustion of phosphorus succeeds equally well in atmospheric air, with only these two differences: (1) the combustion is much less rapid, since it is slowed down by the large proportion of azotic gas which is mixed with oxygen gas; (2) only a fifth of the air at most is consumed, since only the oxygen gas is consumed, and so the proportion of azotic gas becomes so great towards the end of the procedure that combustion can no longer take place.

5.17　　　As I already mentioned, phosphorus, whether it burns in ordinary air or in oxygen gas, is transformed into a very light, flaky material, acquiring completely new properties: though it is initially insoluble in water, it not only becomes soluble, but it draws the moisture from the air with astonishing

---

* For more details see the memoir that Laplace and I jointly presented (*Mémoires de l'Académie*, 1780, p. 355).

5.14 *nothing is emitted from the phosphorus*: If something had been discharged during combustion, it would have shown up as an impurity in the remaining oxygen. But if indeed nothing was emitted, whatever did Lavoisier mean when he spoke of "the vapors emitted" in paragraph 5.12?

It seems clear in retrospect that these "vapors" were the product of combustion in gaseous form (because of the high temperature), which subsequently condensed into "white flakes" as it cooled—leaving pure oxygen as the sole gas in the flask.

5.17 *phosphorus ... is transformed*: The phosphorus that combined with oxygen changed (i) from a cohesive solid to solid flakes, (ii) from being insoluble to highly soluble in water, forming solutions much denser than water, (iii) from a nearly tasteless substance to one whose solution has an "extremely bitter and sharp taste," and (iv) from a combustible to an incombustible substance. The last three of these properties—solubility, sharp taste, and incombustibility—are characteristic of the class of substances called *acids*.

rapidity, and it melts into a liquid much denser than water, and with a much greater specific weight. As phosphorus before its combustion, it was nearly without taste; by joining with oxygen it takes on an extremely bitter and sharp taste. Lastly, it changes from a combustible to an incombustible substance and becomes what is called an "acid."

5.18   As we will soon see, this ability to change from a combustible substance to an acid by the addition of oxygen is a property common to a great number of bodies. In accordance with sound logic, it is indispensable to use the same word for all processes which yield analogous results. This is the only way to simplify the study of the sciences, and it would be impossible to remember all the particulars if no attempt were made to classify them. We will therefore call the conversion of phosphorus into an acid, and, in general, the combination of any combustible body with oxygen, "oxygenation."

5.19   We will likewise adopt the expression "to oxygenate," and I will accordingly say that when phosphorus oxygenates, it is converted into an acid.

5.20   Sulphur is also a combustible body; that is to say, it has the property of decomposing air and of removing oxygen from caloric. Experiments just like those I explained in detail above readily establish this. But I should point out that is impossible to obtain as precise results with sulphur as were obtained with phosphorus by means of the same procedure. This is because the combustion of sulphur creates an acid that is difficult to condense, and because

---

5.20 *a combustible body ... has the property of ... removing oxygen from caloric*: That is, a combustible body has the property of combining with oxygen. Notice this new, implicit definition of "combustion." Far more precise than the casual definition "burning," Lavoisier's characterization identifies the substance-to-substance action that makes combustibility a specifically *chemical* property.

impossible to obtain as precise results with sulphur ... by means of the same procedure: Pay careful attention to the construction of Lavoisier's sentence. He is not saying that precise results cannot be obtained with *sulphur*, only that they cannot be obtained by the *same procedure* as was used for phosphorus. A crucial feature of that procedure was the production of an acid that *condensed into a solid* as it cooled, and so was easy to distinguish from the oxygen gas.

the combustion of sulphur creates an acid that is difficult to condense: If the combustion product remains gaseous, it will be separable from the oxygen only by some additional procedure that was not necessary with phosphorus. (Lavoisier describes methods for separating gases in Part III of the complete *Treatise*.)

sulphur itself is very hard to burn, and because it is soluble in a number of different gases. But I can assert on the basis of my experiments that when sulphur burns, it absorbs air; that the acid sulphur forms is much heavier than sulphur itself; that the weight of this acid is equal to the sum of the weights of the sulphur and the oxygen it has absorbed; and finally, that this acid is heavy, incombustible, and can combine with water in all proportions. The only uncertainty that remains concerns the quantities of sulphur and oxygen that make up this acid.

5.21   Charcoal, which should be considered a simple, combustible substance in light of the foregoing, also possesses the property of decomposing oxygen gas and removing the base from caloric. But the acid that results from this combustion does not condense at the temperatures and pressures in which we live. It remains in a gaseous state, and a great quantity of water is required to absorb it. Moreover, this acid has all the properties common to acids, but to a lesser degree, and like other acids, it unites with all bases capable of forming neutral salts.

5.20, cont. *sulphur itself is very hard to burn*: It not difficult to *ignite* sulphur, as Lavoisier's own experiment shows. But if it does not burn *completely*, the experiment will not yield "precise results."

and because [sulphur] is soluble in a number of different gases: The term "soluble" has a broad range of meanings for Lavoisier, including even *combination* with another substance (compare 1.28, comments). Sulphur does indeed "dissolve" in gases in the sense of passing directly from the solid to the gaseous state—a process that is today called "sublimation" rather than dissolution. But any sulphur lost to sublimation in this experiment would be undetectible by weighing, so it is hard to see why that would interfere with "precise results."

The only uncertainty ... concerns the quantities of sulphur and oxygen: Lavoisier will reveal in Chapter 6 that there are *two* acids of sulfur, and that one of them readily combines with oxygen to produce the other. Thus he has reason to believe that combustion of sulphur results in a *mixture of two acids*; and this would certainly militate against finding a single, precise ratio of sulphur to oxygen.

5.21 *a great quantity of water is required to absorb it*: Charcoal, like phosphorus and sulphur, forms an acid when it combines with oxygen in the procedure Lavoisier is about to describe. The acid so produced is a gas that is soluble in water—but only to a small degree.

all the properties common to acids: These properties were noted in 5.17; but Lavoisier here adds one more, the ability to unite with a number of other substances to form "neutral salts." Lavoisier incorporated a fuller discussion of salts in the *Treatise*; but in the present selection salts will be mentioned only one more time, in Chapter Seven.

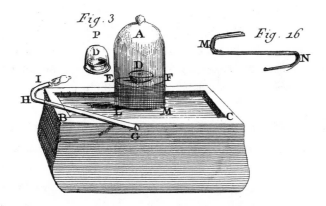

5.22   As in the case of phosphorus, the combustion of charcoal can be produced under a glass bell jar A, Plate IV, Figure 3, filled with oxygen gas and
inverted in mercury. But since the heat of even a red-hot iron wire is insufficient to ignite it, a tiny piece of tinder and a speck of phosphorus should be
placed on top of the charcoal. The red-hot iron easily ignites the phosphorus,
and the fire spreads to the tinder and then to the charcoal.

5.23   The detailed account of this experiment may be found in the memoirs
of the Academy.* This report indicates that it requires 72 parts of oxygen by
weight to saturate 28 parts of charcoal, and that the resulting aeriform acid
has a weight exactly equal to the sum of the weights of the charcoal and oxygen used to produce it. This aeriform acid has been called "fixed air" or "fix'd
air" by the first chemists who discovered it. They did not know at the time
whether this air was similar to the air of the atmosphere or whether it was
another elastic fluid, tainted and corrupted by combustion. But since it is
now certain that this aeriform substance is an acid and that it is formed by the
oxygenation of a base like all other acids, it is easy to see that the name "fixed
air" is inappropriate.

5.24   Once we had burned charcoal in the proper apparatus for determining
the quantity of caloric released, Laplace and I found that burning one livre of

* *Mémoires de l'Académie*, 1781, p. 448.

5.23 *it requires 72 parts of oxygen ... to saturate 28 parts of charcoal*: In keeping
with Lavoisier's earlier verb "absorb," the charcoal is said to be *saturated* with oxygen
when it has absorbed as much oxygen as it possibly can.

oxygenation of a base: Here again, it is important to keep Lavoisier's use of "base"
in mind (1.28, comment).

5.24 *the proper apparatus for determining the quantity of caloric released*: The *ice
calorimeter*, depicted in the comment to paragraph 1.43.

charcoal melts 96 livres 1 ounce of ice and that 2 livres 9 ounces, 1 gros, 10 grains of oxygen combine with charcoal in this process, producing 3 livres 9 ounces, 1 gros, 10 grains of acidic gas. This gas weighs 0.695 grains per cubic inch, which means that the total volume of acidic gas produced by the combustion of a livre of charcoal is 34,242 cubic inches.

5.25   I could provide many additional examples of this kind, and I could present a broad array of facts to show that acids form when substances are oxygenated. But the approach I have determined to take, which consists in proceeding only from what is known to what is unknown and only presenting, to the reader, examples drawn from what I have already explained to him, prevents me from presenting certain facts that I will discuss later. Besides, the three examples I have just cited provide a sufficiently clear and precise idea of the way acids are formed. It is apparent that oxygen is a principle common to all acids, that it is oxygen that constitutes their acidity, and that it is the nature of the substance acidified that distinguishes acids from one another. In each case, then, the acidifiable base (which de Morveau has named the "radical") must be distinguished from the acidifying principle, that is, oxygen.

5.25 *oxygen is a principle common to all acids*: In the course of the *Treatise,* Lavoisier amasses a wealth of evidence for this general principle. But in the next chapter he will have to acknowledge the anomalous *acid obtained from sea salt,* which he is unable either to produce by combustion or decompose to yield oxygen.

it is oxygen that constitutes their acidity: Lavoisier treats this dictum as though it followed from the previous one. But isn't it possible that oxygen is only an incidental, rather than an essential, ingredient in the acids that contain it? Oxygen, we recall, was a ubiquitous component in common air; but Lavoisier didn't assert that oxygen was what constituted air's ability to support combustion until he showed that *pure* oxygen was *surpassingly* supportive of combustion (3.11). But he cannot demonstrate a similarly essential connection with acidity, since oxygen is not acidic on its own.

the radical: From Latin *radix,* root. Thus de Morveau's term "radical" has a meaning essentially the same as Lavoisier's "base."

Chapter Six

On the Nomenclature of Acids in General, and, in Particular, on Those Extracted from Saltpeter and Sea Salt

6.1   Nothing is easier than establishing a systematic nomenclature for acids on the basis of the principles set down in the last chapter. The word "acid" will be the generic name, and then each acid will be differentiated in language as in nature by the name of its base or radical. In general, we will call whatever results from the combustion or oxygenation of phosphorus, sulphur, or charcoal an "acid." We will call the first of these products "phosphoric acid," the second, "sulphuric acid," and the third, "carbonic acid." Likewise, in all such cases we will use the name of the base in order to refer to each acid specifically.

6.2   But there is a remarkable circumstance relating to the oxygenation of combustible bodies, and, speaking generally, to a part of the bodies which transform into acids: they are subject to different degrees of saturation, and the resulting acids, although formed from a combination of the same two substances, have completely different properties, properties that depend on a difference of proportion. Phosphoric acid, and especially sulphuric acid, furnish examples of this. If sulphur is combined with only a little oxygen, it becomes, at this first degree of oxygenation, a volatile acid with a pungent odor and very particular properties. A greater proportion of oxygen converts it into a stable, heavy, odorless acid which, when it combines with other substances, yields products very different from the first. Here it may seem that our nomenclature fails, for it may seem that it would be difficult to derive two names from the same acidifiable base which could express these two degrees of saturation without periphrases or circumlocutions.

6.1 *generic ... specifically*: As Lavoisier urged in his Preliminary Discourse, substances are to be named according to their genus and species, much as plants and animals are named in the system of Linnaeus; see paragraph 0.24 and comment.

carbonic acid: "*Charbon*" is Lavoisier's word for charcoal, "*l'acide carbonique*" the name of its acid.

6.3   But further reflection and, perhaps especially, necessity revealed to us new resources, such that we came to believe that it might be possible for us to express the different varieties of acids by varying the endings of their names. Stahl chose the name "sulphurous acid" for volatile acid of sulphur. We have kept this name, and have given the name "sulphuric" to the acid of sulphur that is completely saturated with oxygen. And so we will say, using this new language, that when sulphur combines with oxygen, it admits of two degrees of saturation: the first is sulphurous acid, which is volatile and pungent; the second is sulphuric acid, which is odorless and stable. We will adopt the same change in endings for every acid that possesses several degrees of saturation. Thus there will be both phosphorous and phosphoric acid, also acetous and acetic acid, and likewise in other cases.

6.4   This whole part of chemistry would have been extremely simple, and the nomenclature for acids would never have presented difficulties, if the radical or acidifiable base of each acid were known at the time the acid was discovered. Phosphoric acid, for example, was discovered only after the discovery of phosphorus, and the name assigned to it was derived, as a result, from the name of the acidifiable base from which it is formed. But, on the contrary, when an acid was discovered before its base, or rather when we had not yet determined the base of the acid at the time of its discovery, then the acid and the base were given names which had no relation to each other. It is not only a burden to keep these useless names in one's memory, they impress beginners' minds and even the minds of the most experienced chemists with false ideas, which only time and reflection can efface.

6.5   The acid of sulphur will serve as an example. This acid was obtained from vitriol of iron in the early years of chemistry, and so it was called "vitriolic acid." This name was borrowed from the name of the substance from which it was extracted. At that time, no one knew that this acid was the same as the one obtained from sulphur by means of combustion.

6.6   The same may be said of the aeriform acid that was originally given the name "fixed air." No one knew that this acid resulted from the combination

---

6.5 *vitriol of iron*: "Vitriols" were so named because, as colored transparent crystals, they were judged to have a *glass-like* appearance (Latin *vitreolum*, little glass). The different vitriols were commonly distinguished by color, as blue vitriol, green vitriol, etc. But since the different varieties were so often found in association with specific metal ores, they came to be designated by those metals as well. Thus, blue vitriol was also known as *vitriol of copper*, while green vitriol was *vitriol of iron*.

of charcoal and oxygen. Because of this, it was given any number of different names, none of which conveyed precise ideas. Revising and correcting the names that were used in the older chemical language for these acids could not have been easier. We have changed the name "vitriolic acid" to "sulphuric acid," and "fixed air" to "carbonic acid." Nevertheless, it was not possible for us to take the same approach when it came to naming acids with unknown bases. In these cases we were forced to take the opposite approach, and instead of deriving the name of the acid from its base, we have, on the contrary, named the base according to the name of the acid. This is the approach we took when we named the acid obtained from sea salt or cooking salt. In order to release this acid, simply pour sulphuric acid on sea salt. A lively effervescence arises immediately, pungent white vapors are emitted, and by applying a gentle heat, all of the acid is released. Since the acid occurs naturally in the state of a gas at the degree of temperature and pressure in which we live, special precautions must be taken to retain it.

6.7   The most convenient and simple apparatus for experiments on a small scale consists of a little retort, G, in which some well-dried sea salt has been placed (Plate V, Figure 5). Concentrated sulphuric acid is then poured over the sea salt, and the mouth of the retort is immediately inserted under small glass vessels or bell jars, A, that were previously filled with mercury. As the

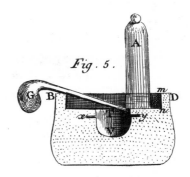

acid gas is released, it passes into the vessel and makes its way to the top, displacing the mercury. When the release of the gas subsides, gentle heat is applied and gradually increased until no more gas appears. This acid has a great affinity for water: the water absorbs an enormous quantity of it. This may be established by adding a thin layer of water to the glass vessel containing the acid: the acid immediately combines with the water, disappearing entirely. We exploit this affinity in the laboratory and in the arts when we produce acid of sea salt in liquid form by means of the apparatus depicted in

---

6.7 *the acid gas*: That "the acid obtained from sea salt" is indeed an *acid* was so universally accepted, Lavoisier evidently saw no need to list its qualifying properties. He does mention *solubility* later in the paragraph; and the solution has a typically acidic "sharp" taste, which he does not mention. But in comparison with other acids, the "acid from sea salt" is decidedly anomalous, as he will shortly discuss.

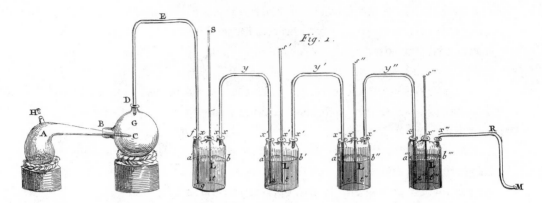

Plate IV, Figure 1. This apparatus consists of (1) a retort, A, in which sea salt has been placed, and in which sulphuric acid is poured by means of the short cylindrical opening, H; (2) the spherical glass vessel, CB, set up to receive the small quantity of liquid released; and (3) the series of bottles with two necks L, L′, L″, L‴ that are half-filled with water. This water absorbs the acid gas that is released during distillation. The apparatus is more fully described in the last part of this work.

6.8   Although we have not yet succeeded either in composing or decomposing the acid obtained from sea salt, nevertheless it is undoubtedly created, like all the others, when an acidifiable base is united with oxygen. We have named this unknown base "muriatic base" or "muriatic radical," taking this name, on the model of Bergman and de Morveau, from the Latin word *muria*, formerly a word for sea salt. Thus, although we have not been able to determine the composition of "muriatic acid" exactly, this term, as we will use it, will refer

---

6.7, cont. *The apparatus is more fully described...*: Lavoisier's more extensive description is not included in the present selection.

6.8 *it is undoubtedly created ... when an acidifiable base is united with oxygen*: What has happened to Lavoisier's noble principle, "one must never infer something that can be ascertained through direct experiment," which he pronounced in paragraph 5.9? For, as he acknowledged at the outset of the present paragraph, his attempts either to *compose the acid from oxygen* or *decompose it to obtain oxygen* have both failed!

   It is too easy to fault Lavoisier on this point. We modern readers enjoy the benefit of having been handed an alternative theory, one that singles out *hydrogen* (see Chapter Eight), not oxygen, as the essential ingredient of acids. But should Lavoisier really have abandoned the powerful and beautiful oxygen theory of acids, in the face of all the evidence favoring it, on the strength of a single counter-example? His attempts to demonstrate composition and decomposition "have not *yet* [*encore*] succeeded"; that need not mean they will *never* succeed. Moreover, he has a very reasonable explanation for both failures; see the next comment.

to a volatile acid which naturally occurs in the state of a gas at the degree of heat and pressure we typically experience. Great quantities of it combine with water very easily, and its acidifiable radical clings with such strength to oxygen that we have not yet discovered a way to separate them.

6.9   If we someday manage to relate the muriatic radical to some known substance, we will then have to change its name, and give it one analogous to the name of the base whose nature was discovered.

6.10   And something else well worth noting occurs with muriatic acid. Like the acid of sulphur and several others, it admits of different degrees of oxygenation. But adding more oxygen to muriatic acid produces an effect completely opposite to the effect produced in the case of acid of sulphur. The first degree of oxygenation transforms sulphur into a volatile, gaseous acid that only mixes with water in small quantities. With Stahl, we designate this acid by the name "sulphurous acid." A greater measure of oxygen changes sulphurous acid into sulphuric acid, that is, into an acid with more conspicuous acidic properties, an acid that is much more stable, can exist in a gaseous state only at a high temperature, has no odor at all, and unites in great quantities with water. Just the opposite occurs in the case of muriatic acid: the addition of oxygen renders it more volatile, with a more penetrating odor. Less of it can be dissolved in water, and its acidic quality diminishes. We had at first attempted to express these two degrees of saturation, as we had done in the case of the acid of sulphur, by varying the endings. We would have named the acid which is less saturated with oxygen "muriatous acid," and the one more saturated "muriatic acid." But this acid possesses such peculiar attributes that nothing like it is known to chemistry, so we deemed it necessary to make an exception. We decided that a satisfactory name for it would be "oxygenated muriatic acid."

[*the muriatic radical*] *clings with such strength to oxygen that we have not yet discovered a way to separate them*: Such a strong coupling, if true, would obviously explain Lavoisier's failure to "decompose" (6.8) muriatic acid, that is, to release oxygen from it. It would also explain his failure to "compose" (6.8) the acid by oxygenating its radical; since in order to obtain that radical, it would first have to be disengaged from oxygen. But the difficulty has all the appearance of one that must eventually yield to improved techniques. Lavoisier can be excused for not giving up easily.

6.10 *Just the opposite occurs in the case of muriatic acid*: When muriatic acid is oxygenated, its acidic properties become less, not more, pronounced.

oxygenated muriatic acid: For all other acids, the name ending in *-ous* denotes both the lesser degree of oxygenation and the less clearly acidic form, the *-ic* ending signifying the greater degree of oxygenation and the more definitively acidic substance; so those adjectival endings would be misleading if they were applied to the muriatic radical.

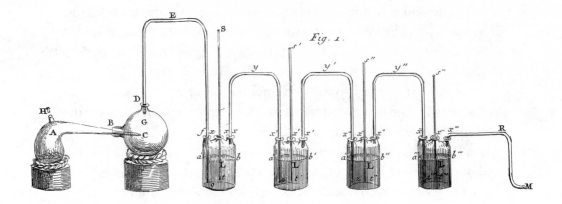

6.11   There is another acid that we decided to name in the same way as we named muriatic acid, although its base is better known. This is the acid that chemists up until now have referred to as "nitrous acid." This acid is obtained from nitre or saltpeter by a process analogous to the one used to obtain muriatic acid. It is separated from the base to which it is attached in the same manner, by means of sulphuric acid, and, as before, with the apparatus depicted in Plate IV, Figure 1. As the acid passes through the apparatus, part of it condenses in the glass flask while the rest is absorbed by the water in bottles L, L′, L″, L‴ which becomes green at first, then blue, and then finally yellow, according to the degree of the acid's concentration. During this process, a large quantity of oxygen gas mixed with a small amount of azotic gas is emitted.

6.12   Like all acids, the acid extracted from saltpeter is composed of oxygen united to an acidifiable base, and in fact, the first sound demonstrations of the existence of oxygen in acids were made using this acid. The two principles that constitute it do not have a strong hold on one another, and so they can be easily separated by introducing to oxygen a substance with which it has a greater affinity than it has with the acidifiable base that constitutes the acid of nitre. Experiments like these led to the discovery that nitrous acid contains azote, the base of noxious gas, which is its acidifiable base. Azote is

---

6.11 *nitrous acid*: The traditional name "nitrous" was a simple adjective denoting the acid's source (nitre); its *-ous* ending did not have the systematic significance Lavoisier bestows upon *-ous* and *-ic*.

     nitre or saltpeter: Known by both names, this substance is found as crystalline deposits on cave walls; it may also be extracted from guano deposits, using water.

6.12 *the first sound demonstrations of the existence of oxygen in acids were made using this acid*: The demonstrations were conducted by Lavoisier himself in 1776. He heated the acid with mercury; at the end of a lengthy process, the mercury was recovered intact along with a mixture of gases, one of them oxygen.

therefore the true nitric radical, or, in other words, the acid of nitre is a true azotic acid. So it seemed that in order to remain consistent with ourselves and with our principles, we would have to adopt one of these two ways of expressing ourselves.

6.13   However, other considerations deterred us. First of all, it seemed difficult to us to change the names "nitre" and "saltpeter," which are used throughout the arts, in society, and in chemistry. Furthermore, we thought that we should not call azote the "nitric radical," because azote is also the base of volatile alkali (ammoniac), as Berthollet has discovered. We will therefore continue to refer to the base of the non-respirable part of atmospheric air as "azote," which is at the same time the nitric radical and the ammoniac radical. We will also retain the words "nitrous" and "nitric" for the acid drawn from nitre or

---

nitrous acid contains azote ... which is its acidifiable base: That nitrous acid "contains azote" is no more than what Lavoisier acknowledged earlier, in paragraph 4.10. But that was before he determined that every acid is distinguished by a corresponding acidifiable base (5.25). Now that the general constitution of acids is established, he is able to report that azote is not merely *a* constituent; it is *the* specific acidifiable base of nitrous acid.

it seemed that ... we would have to adopt one of these two ways of expressing ourselves: The two alternatives are (i) "azote is the true nitric radical" and (ii) "the acid of nitre is azotic acid." The first locution would relinquish the name azote. Lavoisier might have taken that path in Chapter Four: instead of styling the non-respirable part of air *azote*, he might have called it the nitric radical, reflecting its status as the base of the acid of nitre. The second locution would abandon the name nitre, replacing it with a term that derives from *azote*, and so renaming "acid of nitre" as "azotic acid." But "other considerations deterred" him from doing either.

6.13 *it seemed difficult to us to change the names...*: He did, indeed, state in the Preliminary Discourse that "it was our aim to retain all common or customary names" (0.22).

Furthermore, we thought that we should not call azote the "nitric radical," because azote is also the base of volatile alkali...: Lavoisier's implicit argument here is that we have no way to decide which property is more fundamental—the ability to form nitre or the ability to form volatile alkali. His reason for rejecting the term "nitric radical" here is the same as his reason for declining "nitrogenic radical" in paragraph 4.10.

We will also retain the words "nitrous" and "nitric" for the acid drawn from nitre: Nevertheless, he will hold out for one innovation, the systematic use of *-ous* and *-ic* to indicate degrees of oxygenation. Lavoisier will shortly identify two kinds of "acid of nitre"; but for many years the traditional name "nitrous acid" was used indiscriminately for both.

saltpeter. Several chemists of great authority have disapproved of the way we have stooped to these older names. They would have preferred us to have focused all of our efforts on the perfection of nomenclature, to have reconstructed the edifice of chemical language from top to bottom without troubling ourselves to make connections with former usages, the memory of which time will gradually obliterate. And so we have found ourselves subjected to the criticisms and complaints of the two opposing sides simultaneously.

6.14   The acid of nitre can occur in many different states depending on its degree of oxygenation, that is to say, on the proportions of azote and oxygen which compose it. A first degree of oxygenation of azote yields a particular gas, which we will continue to call "nitrous gas." It is composed of approximately 2 parts oxygen and one part azote by weight, and in this state it does not mix with water. The azote in this gas, far from being saturated with oxygen, retains a great affinity for it, and attracts it with such vigor that it draws oxygen away from any atmospheric air which comes in contact with it. In fact, we now combine nitrous gas with atmospheric air in order to determine the quantity of oxygen the air contains and to gauge its wholesomeness. This addition of oxygen changes nitrous gas into a powerful acid that has a great affinity for water, and which is itself capable of different degrees of oxygenation. If

---

6.13, cont. *Several chemists of great authority have disapproved*: They were evidently not as hesitant about the ambivalent choice of names as Lavoisier was. As early as 1791 the distinguished chemist Jean-Antoine Chaptal proposed replacing the name "azote" by "nitrogen." That name, he wrote, "is deduced from the characteristic and exclusive property of this gas, which forms the radical of nitric acid." Modern chemistry has unswervingly adopted this path.

6.14 *A first degree of oxygenation of azote yields ... nitrous gas*: Notice that this lowest degree of oxygenation—two parts oxygen to one of azote—is not even called *acid*.

In fact, we now combine nitrous gas with atmospheric air in order to determine the quantity of oxygen the air contains: This was the "nitrous air test" developed by Joseph Priestley. He would mix a sample of common air with nitrous gas confined over mercury. The nitrous gas would snatch oxygen from the sample, and the volume of the mixture would decrease accordingly. The more oxygen in the sample, the greater the decrease—and the more "wholesome" was the sample judged to be.

addition of oxygen changes nitrous gas into a powerful acid: This represents a second degree of oxidation of azote, and the lowest degree that produces an acid. Lavoisier will call it "nitrous acid" later in the paragraph.

the proportion of oxygen to azote is below three parts to one, the acid is red and fumes profusely. In this state we call it "nitrous acid." This acid releases nitrous gas when it is gently heated. Four parts of oxygen to one part azote produces a white, colorless acid, which is more stable when exposed to heat than nitrous acid, and less odorous, and whose two constitutive principles are more strongly united to one another. In accordance with the principles outlined above, we have given it the name "nitric acid."

6.15   Thus, of these two acids, nitric acid is the acid of nitre with the greater proportion of oxygen, and nitrous acid is the acid of nitre with the greater proportion of azote (or, what amounts to the same thing, with the greater proportion of nitrous gas). Finally, nitrous gas is azote that is not sufficiently saturated with oxygen to possess the properties of acids. Later we will call this an "oxide."

If the proportion of oxygen to azote is below three parts to one, ... we call it "nitrous acid": The proportion Lavoisier cites is by volume, not weight.

Four parts oxygen to one part azote produces ... "nitric acid": This represents the third degree of oxygenation of azote, and the second that produces an acid.

6.15 *...not sufficiently saturated with oxygen to possess the properties of acids. Later we will call this an "oxide"*: Oxides will be the principal topic of the next chapter.

Chapter Seven

On the Decomposition of Oxygen Gas by Metals,
and on the Formation of Metallic Oxides

7.1   When metallic substances are heated to a certain temperature, oxygen has more affinity for them than for caloric. Because of this, all metallic substances, except gold, silver and platinum, have the property of decomposing oxygen gas, taking hold of its base, and disengaging caloric. We have already seen above how this decomposition of air by mercury and iron takes place, and we have observed that the decomposition by mercury can only be considered a slow combustion, whereas decomposition by iron, on the contrary, is very quick and is accompanied by a brilliant flame. If a certain degree of heat is required for these procedures, this is only in order to separate the particles of the metal from one another so as to diminish their affinity of aggregation, or, what is the same thing, the attraction that they exert on one another.

7.2   Metallic substances, during calcination, increase in weight in proportion to the oxygen that they absorb, and at the same time they lose their metallic sheen and are reduced to an earthy powder. Metals in this state ought not to be considered completely saturated with oxygen, since the effect of metals on oxygen is balanced by the force of attraction that caloric exerts on oxygen.

7.1 *we have observed that the decomposition* [*of oxygen*] *by mercury can only be considered a slow combustion*: Here, as previously, "decomposition" of a gas means releasing the gas's base from caloric. Mercury accomplishes this by combining with the base. Lavoisier has not explicitly affirmed that calcination of mercury is "slow combustion"; but the interpretation is inescapable, since calcination and combustion are both processes of *union with oxygen*.

7.2 *Metals in this state...*: That is, the state of having combined with oxygen.

...ought not to be considered completely saturated with oxygen: Why is Lavoisier so quick to caution readers against this supposition? Possibly because when he calcined mercury and iron in Chapter Three, nothing happened to suggest that these metals were capable of combining with a greater quantity of oxygen than they actually did. It would be natural, then, to entertain the possibility that these metals were capable of only a single state of oxidation—which would necessarily be a state of saturation.

the effect of metals on oxygen is opposed by the ... attraction that caloric exerts on oxygen: Lavoisier previously called attention to such a rivalry in paragraph 3.14.

Thus, in the calcination of metals, oxygen is actually subject to two forces, the force exerted by the caloric, and the force exerted by the metal, and it only tends to unite with the metal because of the difference between these two forces, that is, as a result of the extent to which the force exerted by the metal exceeds the force exerted by the caloric. And in general this excess in not very significant. So metallic substances, unlike sulphur, phosphorus, and charcoal, are by no means converted into acids when they oxygenate in the air or in oxygen gas. They become, rather, intermediary substances: they begin to approach a saline state without acquiring every saline property.

7.3   Previous chemists used the term "calx" to refer not only to any metal that had been brought to this state, but to any substance that did not melt after long exposure to fire. In this way, they used "calx" as a generic term, under which they jumbled together limestone with metals; that is to say, they jumbled together a substance which, in calcination, is converted from a neutral salt to an earthy alkali and loses half its weight, with sub-stances which, in calcination, gain sometimes more than half their weight

7.4 and become almost acids. It would have been contrary to our principles to

---

So metallic substances ... are by no means converted into acids when they oxygenate: He seems to be of the opinion that metals simply cannot exert a strong enough hold upon the oxygen they combine with to give them acidic properties.

they begin to approach a saline state: A "saline" state (from Latin *sal*, salt) is one that resembles, or contains, salt. In later chapters, not part of the present selection, Lavoisier notes that oxidized metals become salts *when acted upon by acids*. Lavoisier terms them the "salifiable bases" of their respective salts; and he labels acids as "the true salifying principles." He also acknowledges *earths* and *alkalis* as two other classes of salifiable bases. We can perhaps begin to see how the term "base" could gradually develop its modern connotation as the counterpart to "acid."

7.3 *calx*: Lavoisier has been scrupulous in using this term exclusively to denote a metal that has been heated sufficiently to combine with oxygen and so gain weight. But since others have used "calx" so loosely, he will in paragraph 7.4 adopt the term "oxide" instead.

a substance which, in calcination, is converted ... to an earthy alkali: The "earthy alkali" he has in mind here is commonly called *quicklime*. Previous chemists were wont to call quicklime a "calx" because it was produced by *calcination* of limestone.

substances which, in calcination, ... become almost acids: When substances are calcined they unite with oxygen; and oxygenation produces either an acid (5.25) or a state that is on the way to becoming an acid (6.15). So by calling all products of calcination "calces," previous chemists lumped alkaline and acidic substances into the same class, even though their properties are very different.

classify under the same name substances so different and, above all, to keep a name for metals so likely to give rise to false ideas. Because of this we have banned the term "metallic calces" and have substituted in its place the term "oxides" from the Greek ὀξύς.

7.5   From this it is clear how fruitful and expressive is the language we have adopted: a first degree of oxygenation forms oxides; a second degree forms acids with names ending in "ous," such as nitrous acid, and sulphurous acid; a third degree forms acids with names ending in "ic" such as nitric and sulphuric acids, and, finally, we can express a fourth degree of oxygenation of these substances by adding the qualifying adjective "oxygenated," as we noted in the case of oxygenated muriatic acid.

7.6   And by the term "oxides" we refer not only to those oxygen compounds involving metals. For we saw no reason not to use this term to express the first degree of oxygenation of all substances which [when so oxygenated] do not become acids but, rather, approach a saline state. Thus we will call sulphur, when it first begins to burn and becomes soft, "oxide of sulphur," and we will refer to the yellow substance that remains after phosphorus burns "oxide of phosphorus."

7.7   We will also call nitrous gas, which is the first degree of oxygenation of azote, "oxide of azote." And, finally, there will be oxides in the animal and vegetable kingdoms; and I will show in what follows how this new language illuminates every operation of art and nature.

7.8   As we have seen, almost every metallic oxide has its own particular color. And these colors vary not only from metal to metal but even among the different degrees of oxygenation of the same metal. We thus found ourselves obliged to add to each oxide two epithets, one indicating the metal oxidized, and one indicating its color. Thus we will use the terms "black iron oxide," "red iron oxide," and "yellow iron oxide" to refer to the substances previously known as "martial ethiops," "colcothar," and "iron rust" (or "ochre") respectively.

7.9   In the same way we will speak of "grey lead oxide," "yellow lead oxide," and "red lead oxide," and these expressions will designate the substances formerly known as "lead ash," "massicot" and "minium" respectively.

7.4 *we have banned the term "metallic calces" and have substituted ... the term "oxides"*: Even if the common usage of "calx" had not robbed the term of precision, it would still fail to express the *composition* of its referent (0.28), conveying instead merely the method for producing it. "Oxide" indicates combination with oxygen— an attribute that penetrates far more deeply into the nature of the substance.

7.10   The new names will sometimes be rather long, especially when it is necessary to express whether the metal has been oxidized in the air, by detonation with nitre, or through the action of acids, but at least they will always be just and will give rise to the precise idea of the object which corresponds to them.

7.11   The tables appended to this treatise will make all of this clearer.

7.10 *detonation with nitre*: A *detonation* is an explosion accompanied by a sudden loud report; from Latin *tonare*, thunder. In Part II of this work Lavoisier writes: "One can also oxygenate combustible substances by combining them ... with nitrate of potash [nitre]. At a certain degree of heat, the oxygen leaves the nitrate ... in order to combine with the combustible body. However, these sorts of oxygenation must be attempted with extreme precaution and with very small quantities. Oxygen combines with nitrates with [an] immense quantity of caloric [which] becomes suddenly free at the moment of its combination with the combustible bodies, resulting in dreadful detonations which nothing can withstand."

through the action of acids: Acids typically act slowly on most metals, although zinc and some others dissolve quickly in acids, producing an effusion of bubbles ("effervescence"). In contrast, acids act rapidly, without effervescence, on oxidized metals. In both cases, a salt is produced. So when Lavoisier speaks of a metal being *oxidized* by action of an acid, he means the production of a *salt*, not a *metal oxide*: if an oxide were produced, it would immediately react with the acid to form the salt (7.2, comments). Lavoisier regards the metal constituent of a salt as "oxidized" insofar as both it and oxygen, being constituents of the same substance, are in combination with one another.

7.11 *The tables appended to this treatise will make all of this clearer*: Notwithstanding Lavoisier's inviting promise, his tables could not be included in the present selection.

Chapter Eight

On the Radical Principle of Water, and on the Decomposition of Water
by Charcoal and by Iron

8.1   Until very recently, water was considered a simple substance, and our predecessors had no qualms calling it an element. They never doubted that it was an elementary substance since they had not yet reached the point of decomposing it or, rather, because the decompositions of water which took place before their eyes escaped their notice. But we shall soon see that water is no longer an element for us. I will not give an account of the history of this discovery here, which is very recent and still controversial.*

8.2   I will content myself, instead, with reporting the main proofs of the decomposition and recomposition of water. I will go so far as to say that when these proofs are weighed without prejudice they will be found to be conclusive.

First experiment.
Set-up

8.3   Take a glass tube, EF, with a diameter of 8 to 12 lines, and place it over a furnace at a slight incline from E to F (Plate VII, Figure 11). Attach a small glass retort, A, filled with common distilled water to the upper end of the tube, E, and to the lower end, F, attach a spiral tube, SS′, which has been attached at its other end, S′, to one mouth of a double-mouthed flask, H. And, finally, attach to H's other mouth, a curved glass tube, KK, designed to conduct aeriform fluids, or gases, into a device appropriate for determining their quantity and quality.

8.4   In order to ensure that the experiment succeeds, the tube EF should be made from well-annealed green glass which is difficult to melt. Coat the outside with a lute of clay mixed with a cement made of pulverized stoneware pottery, and in case the glass should bend as it melts, it should be supported in the middle with an iron bar across the furnace. Porcelain pipes are preferable to glass pipes, but it is difficult to obtain porcelain pipes which are not porous. These pipes are almost always found to have some holes that allow air or vapors to pass through.

---

* For this, see *Mémoires de l'Académie des Sciences*, 1781.

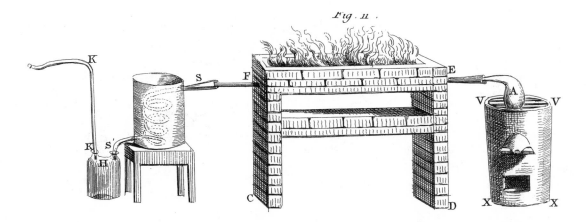

Fig. 11.

8.5   When everything has been set up as indicated, light a fire in the furnace, EFCD, and stoke it in such a way as to make the tube EF glow red without melting it. At the same time, light a fire in the furnace, VVXX, so as to keep the water in retort A boiling.

Result

8.6   As the water in the retort boils and evaporates, it fills the interior of the tube EF, and expels the common air which leaves by the tube KK. Aqueous gas then condenses as it cools in the spiral tube SS′ and water falls, drop by drop, into the flask H.

8.7   When this procedure is continued until all the water in the retort is evaporated and the vessels are allowed to drain completely, flask H is found to contain a quantity of water exactly equal to that which was in the retort A, without any gas having been released. We have thus ensured that this procedure is nothing but simple, ordinary distillation, and the outcome is absolutely the same as it would have been if the water had not been brought to an incandescent state as it passed through the intermediary tube EF.

Second experiment.
Set-up

8.8   Set everything up as in the previous experiment, with one difference: place 28 grains of charcoal in the tube EF. The charcoal should be crushed into

---

8.6 *aqueous gas*: That is, water in the gaseous state (4.7). It condenses into liquid water.

8.7 *an incandescent state*: Here, as also in the following paragraph, "incandescent" means *intensely hot*, rather than its more literal signification, "glowing." Compare Lavoisier's expression "red heat" in paragraph 8.14, where the *container*, not the substance being heated, becomes incandescent.

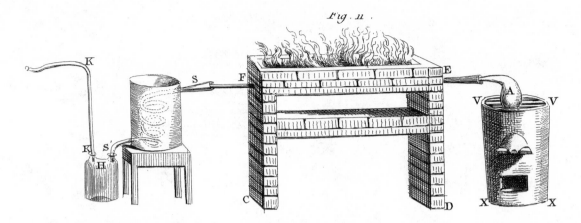

Fig. 11.

middle-sized pieces and exposed beforehand to incandescent heat in closed vessels. Then boil water in retort A until it completely evaporates as before.

Result

8.9   Water from the retort is distilled in this experiment as in the preceding experiment. It condenses in the spiral tube and drips drop by drop into flask H. But at the same time it releases a considerable quantity of gas, which escapes by tube KK, and is collected in an appropriate receptacle.

8.10   At the end of the process, only a few specks of ash remain in tube EF: the 28 grains of charcoal have completely disappeared.

8.11   When the gases released are examined carefully, they are found to weigh a total of $113\frac{7}{10}$ grains.* There are two types of gas: 144 cubic inches of carbonic acid gas, weighing 100 grains, and 380 cubic inches of an extremely light gas which weighs $13\frac{7}{10}$ grains and ignites in the presence of a flame when it is in contact with the air. If the weight of the water that has passed into the flask is checked, it is found to have diminished by $85\frac{7}{10}$ grains.

> * The third part of this work includes a detailed account of the method for separating different types of gas and for weighing them.

8.8 *exposed beforehand to incandescent heat in closed vessels*: This process drives off any volatile impurities. Such a degree of heat might trigger calcination or combustion of the charcoal (5.21–5.22), although the limited air supply in "closed vessels" helps limit the extent to which this can occur. But since any combination with oxygen will diminish the supply of solid charcoal, the heating is to take place "beforehand"—that is, before weighing.

8.11 *two types of gas*: From Chapter Five, we know that the 100 grains of carbonic acid gas represents the combination of 28 grams charcoal and 72 grams oxygen (5.23).

   extremely light: Since 380 cubic inches weigh $13\frac{7}{10}$ grains, one cubic inch weighs .036 grains, about one-fourteenth the weight of a cubic inch of oxygen (4.8).

8.12   Thus, in this experiment, $85\frac{7}{10}$ grains of water plus 28 grains of charcoal has formed 100 grains of carbonic acid plus $13\frac{7}{10}$ grains of a certain flammable gas. But I have shown above, that in order to form 100 grains of carbonic acid gas, it is necessary to combine 72 grains of oxygen with 28 grains of charcoal. Thus the 28 grains of charcoal placed in the tube removed 72 grains of oxygen from the water, which means that $85\frac{7}{10}$ grains of water is composed of 72 grains of oxygen and $13\frac{7}{10}$ grains of flammable gas. We shall soon see that we should not imagine that this gas was released from the charcoal, and that it is consequently a product of water.

8.13   In my account of this experiment, I have suppressed certain details that would only tend to complicate it and obscure readers' ideas. The flammable gas, for example, dissolves a little of the charcoal, and this factor increases its weight and on the contrary diminishes the weight of the carbonic acid. The resulting differences in the quantities are not very considerable, but I thought I should recalculate them and present the experiment in its complete simplicity, as if no charcoal had dissolved, in order to present the experiment in as simple a form as possible. Moreover, if any suspicions obscure the truth of these conclusions, they will be dispelled by the other experiments I will relate in support of them.

Third experiment.
Set-up

8.14   Arrange all of the equipment in the same way as in the preceding experiment, except instead of 28 grains of charcoal, place 274 grains of soft iron strips shaped into spirals in the tube, EF. Bring the tube to a red heat as in the preceding experiment, and light the fire under the retort, A, keeping the

---

8.12 *I have shown above*: In paragraph (5.23).

85.7 grains of water is composed of 72 grains of oxygen and $13\frac{7}{10}$ grains of a certain flammable gas: He thus obtains 5.25 *to* 1 as the ratio, by weight, of oxygen to flammable gas.

We shall soon see that we should not imagine that this gas was released from the charcoal: This will become apparent in the Third Experiment, especially paragraphs 8.15 and 8.16.

8.13 *The flammable gas ... dissolves a little of the charcoal*: Here, again, "dissolves" probably does not refer to *sublimation* (5.20, comment), since that would not diminish charcoal's availability to form carbonic acid. But at temperatures well within reach of Lavoisier's furnace EFCD, a small amount of charcoal might *combine* with the flammable gas to form another gas that would later be called "methane."

recalculate ... as if no charcoal had dissolved: Lavoisier does not tell us what method he uses to make this correction.

water at a constant boil until it has completely evaporated and all of it has passed through the tube EF and condensed in flask H.

Result

8.15   No carbonic acid gas is released in this experiment. The only gas released is flammable and 13 times lighter than atmospheric air. Its total weight is 15 grains, and it has a volume of around 416 cubic inches. If the initial quantity of the water used in this experiment is compared to the water remaining in flask H, it is found to have lost 100 grains. On the other hand, the iron contained in tube EF, which initially weighed 274 grains, gains 85 grains, and its volume increases considerably. This iron is almost no longer attracted by the magnet, and it dissolves in acids without effervescence. In a word, it becomes a black oxide, precisely as if it had burned in oxygen gas.

Reflections

8.16   The result of this experiment is true oxidation of iron by water, oxidation exactly like that which occurs in the air with the aid of heat. 100 grains of water has been decomposed, 85 grains of oxygen has been united to the iron so as to constitute a black oxide, and 15 grains of a certain flammable gas has been released. Thus, water is composed of oxygen and the base of a flammable gas in the ratio of 85 parts to 15.

8.17   Thus, water contains not only oxygen, which is one of its principles and which it shares in common with many other substances, but another principle which is proper to it, its constitutive radical, and which we were obliged to name. No name seemed more appropriate than "hydrogen," that is to say

---

8.15 *The only gas released is flammable and 13 times lighter than atmospheric air*: This is clearly the same gas as that produced in the second experiment.

the iron ... gains 85 grains: Since its initial weight was 274 grains, it increased in weight by 31 percent—in decent agreement with the 35 percent weight increase when he oxidized iron in Chapter Three (3.24).

it dissolves in acids without effervescence: As noted in the comment to paragraph 7.10, this is characteristic of metal oxides.

In a word, it becomes a black oxide, ... as if it had burned in oxygen gas: This black oxide of iron is identical to the *martial ethiops* produced earlier (3.24). The iron that formed it must, therefore, have united with oxygen and only oxygen; and the only possible source for that oxygen is the water. Then the "flammable gas" must represent a hitherto unknown constituent of water that remains behind when the oxygen is removed.

8.16 *water is composed of oxygen and the base of a flammable gas in the ratio of 85 parts to 15*: That is, a ratio of 5.7 to 1—a value reasonably consistent with the ratio of weights obtained in the previous experiment, 5.25 to 1 (8.12, comment).

the "generating principle of water" from ὕδωρ ("water") and γείνομαι ("I generate"). We will call "hydrogen gas" the combination of this principle with caloric, and the word "hydrogen" will only express the base of this same gas, the radical of water.*

8.18   We have discovered, then, a new kind of combustible body, that is to say, a body with enough affinity for oxygen to remove its caloric from it and to decompose air or oxygen gas. This combustible body has, itself, so much affinity for caloric, that unless it is engaged in a compound, it is always in an aeriform or gaseous state at the ordinary pressures and temperatures in which we live. In this gaseous state it is about 13 times lighter than atmospheric air, water does not absorb it (though it can dissolve a small quantity of water), and it is not fit for the respiration of animals.

8.19   Since the property of burning and flaring up is, for this gas and for all other combustible gases, nothing but the property of decomposing the air and removing oxygen from caloric, it is clear that the gas cannot burn unless it is in contact with air or oxygen gas. Thus, when a bottle is filled with this gas and the gas is lit, it burns gently at the mouth of the bottle and only afterwards inside as the air outside makes its way in. But the combustion is successive and slow: it only occurs at the point of contact between the two gases. It is not the same when one mixes the two airs together before lighting them. If, for example, a burning body (such as a piece of paper or a candle) is brought close to the mouth of a narrow-neck bottle filled with one part oxygen gas and two parts hydrogen gas, the combustion of the two gases occurs instantaneously and with a strong explosion. This experiment should only be made with a very strong green bottle with a volume of no more than a pint and which has been covered with a rag, since otherwise disastrous mishaps can occur: if the bottle ruptures, fragments of glass can shoot out to great distances.

8.20   If all that I have set out concerning the decomposition of water is accurate and true—if this substance is really composed, as I have attempted to establish, of a principle (hydrogen) proper to it, combined with oxygen— it should be the case that by uniting these two principles, one ought to be

---

* I have been bitterly criticized for this expression, "hydrogen," because it has been claimed that that term signifies not "the generating principle of water" but "the offspring of water." But it hardly matters, since either interpretation is appropriate. The experiments described in this chapter prove that water, when decomposed, "gives birth" to hydrogen, and moreover hydrogen "gives birth" to water when it combines with oxygen. One could just as well say that hydrogen "generates" water as that water "generates" hydrogen. [Lavoisier added this note in the 1793 edition. —Tr.]

8.19 *the combustion is successive*: It takes place gradually, not all at once as in the *explosion* mentioned later in the paragraph.

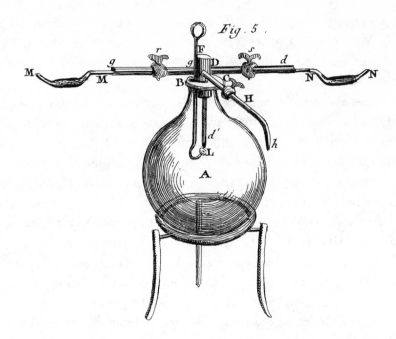

able to create water again. The following experiment demonstrates that this is exactly what happens.

Fourth experiment: Recomposition of water.
Set-up

8.21   Take a spherical crystal flask, with a large opening, and a volume of around 30 pints, and attach a copper plate to it with putty (Plate IV, Figure 5). Four tubes should pass through holes pierced in the plate.

8.22   The first tube, Hh, is designed to be attached at one end, h, to an air pump which can be used to produce a vacuum in the flask. A second tube, gg, is attached at one end, MM, to a reservoir of oxygen gas, and is designed to conduct the gas into the spherical flask. The third, dDd' is attached at one end, dNN, to a reservoir of hydrogen gas. There should be such a small opening at the end of this tube that a very fine needle can hardly pass through it. The hydrogen gas in the reservoir will escape through this tiny opening, and in order to ensure that it escapes quickly enough, it ought to be subjected to a pressure of 1 or 2 inches of water. Finally, the fourth hole in plate BC is fitted with a glass tube which is attached to the plate with putty, and through which passes a metal wire, GL. A little ball is designed to be attached to the end L of the wire which makes it possible to draw off electric sparks from L to d' and, as will soon be apparent, ignite the hydrogen gas. The metal wire is free to move in the tube so that the ball can be separated from the end d' of the tube Dd'. The three tubes dDd', gg, Hh are all fitted with stopcocks.

8.23   The hydrogen and oxygen gases should arrive very dry from their respective tubes that bring them to the flask A, and they should have as much water as possible removed from them. Therefore, make them pass through tubes MM and NN about an inch in diameter and filled with exceptionally deliquescent salt—salt, that is, which sucks up the moisture of the air voraciously, such as acetate of potash, or muriate or nitrate of lime. For the purposes of this experiment, these salts ought to be ground to a coarse powder, so that they do not cake, and so that the gas can pass easily through the interstices between the particles. We will examine the composition of these salts in the second part of this work.

8.24   Make sure in advance that there is a sufficient supply of very pure oxygen gas, and in order to make certain that this gas does not contain any carbonic acid, allow the gas to remain in contact with potash that has been dissolved in water and from which all carbonic acid has been removed by means of lime. Details for obtaining this alkali are provided below.

8.25   The same care should be taken in preparing a double portion of hydrogen gas. The surest method for obtaining the gas unmixed consists in extracting it in the decomposition of water with very pliant and very pure iron.

8.26   When the two gases have been thus prepared, attach the air pump to the tube Hh, and make a vacuum in the large spherical flask A. Introduce one of the two gases, preferably the oxygen first, by means of the tube gg. Then, using a certain degree of pressure, introduce the hydrogen through the tube dDd′,

---

8.23 *the second part of this work*: Not included in the present selection, Part II of the *Treatise* is titled "On the Compounds of Acids with Salifiable Bases, and on the Formation of Neutral Salts."

8.24 *Details for obtaining this alkali*: That is, for obtaining potash. But the procedure is not included in this selection.

8.25 *a double portion of hydrogen*: Lavoisier means this merely as a practical hint. Having performed the experiment previously, he knows that unless he starts with a double volume, he will run out of hydrogen before he consumes the oxygen. But in 1804, Joseph-Louis Gay-Lussac will discover that the 2:1 volume ratio for producing water is *exact*. That finding, together with the now-accepted densities of hydrogen and oxygen, implies that the ratio of oxygen to hydrogen in water *by weight* is 8 to 1, rather than the ratios Lavoisier obtains in paragraphs 8.12, 8.16, and 8.28.

     extracting it from the decomposition of water with very pliant and very pure iron: Ideally, the Third Experiment produces hydrogen, unmixed with carbonic acid gas (8.15); but that acid will be formed if the iron used contains carbon (Lavoisier's "charcoal"), since any carbon will oxidize. *Pliant* iron is soft and bendable—a sign that it contains little or no carbon. (Iron with an admixture of carbon up to about 2% is *steel*, which is not pliant but rigid.) Compare paragraph 3.28.

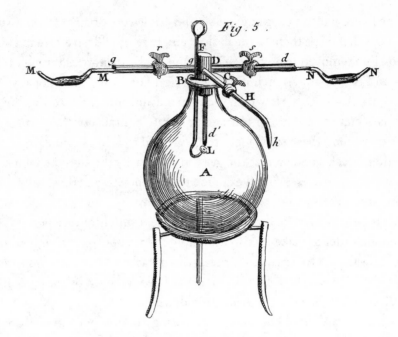

whose extremity d' terminates in a point. Finally, ignite the gas with the help of an electric spark. By letting in a little of both gases into the flask in this way, it is possible to continue the combustion for a long time. I have provided elsewhere a description of the equipment that I used for this experiment and have explained how it is possible to measure with rigorous precision the amount of the two gases consumed. See the third part of this work.

Result

8.27   As the combustion proceeds, water is deposited on the interior walls of

---

8.26 *the equipment that I used*: Above all, the appliance called the *gasometer*, illustrated here. This gargantuan apparatus was capable of delivering a measured quantity of gas at a controlled pressure. In this experiment it provided a constant flow of gas to the combustion flask depicted in Figure 5.

See the third part of this work: Part III of the *Treatise*, not included in the present selection, is Lavoisier's fascinating review of the chemical instruments and operations employed in his investigations.

the flask. The quantity of this water increases bit by bit, and it collects in large droplets which run together at the bottom of the vessel.

8.28   By weighing the flask before and after the procedure, it is easy to determine the amount of water that forms. Thus the experiment provides us with a double verification: for it is possible to verify both the weight of gas employed and the weight of the water formed. These two quantities ought to be equal. By means of an experiment like this one, Meusnier and I found that it takes 85 parts by weight of oxygen and 15 parts by weight of hydrogen to create 100 parts of water. This experiment, which has still not been published, was performed in the presence of a large committee from the Academy. We attended to the experiment meticulously and have reason to believe that the ratio is exact within no more than 2 percent.

8.29   And so, whether from the perspective of decomposition or recomposition, we can consider the following facts to be as invariable and as well-established as any facts can be in chemistry and natural philosophy: that water is not a simple substance; that it is composed of two principles, oxygen and hydrogen; and that these two principles, when separated from one another, have such an affinity for caloric that they can only exist in the form of gases at what are, for us, ordinary temperatures and pressures.

8.30   This phenomenon of decomposition and recomposition of water occurs constantly before our eyes at atmospheric temperature due to the compounded affinities of the substances involved. At least to some extent,

---

8.28 *These two quantities ought to be equal*: That is, the total weight of the two gases, hydrogen and oxygen, should equal the weight of water produced. Here again, as in paragraph 5.7, Lavoisier is tacitly employing the principle of constancy of weight.

  85 parts by weight of oxygen and 15 parts by weight of hydrogen: Exactly the same ratio, this time for *composition* of water, as he found by *decomposition* in the Third Experiment. Notwithstanding Lavoisier's confident estimate of accuracy within 2 percent, it is very far from the 8 to 1 ratio which later investigators will determine and which is still the accepted value today (8.25, comment). But no one disparages Lavoisier's luminous demonstration that water is not an element.

8.29 *water is not a simple substance*: That water is an element was of course a tenet of Aristotelian and other ancient schools of natural philosophy. But the true importance of Lavoisier's finding has little to do with confutation of the ancients. Rather it illustrates what Lavoisier has already expressed in his Preliminary Discourse: that an element or simple substance is nothing more than one that we have so far been unable to decompose—and that even substances long accepted as "elements" may at some point prove to be compound in nature.

phenomena such as the fermentation of spirits, putrefaction, and even the growth of plants depend on the decomposition of water, as will soon be apparent. So it is quite extraordinary that this decomposition escaped the watchful eyes of natural philosophers and chemists until recently, and this suggests that in science, as in ethics, it is difficult to overcome deeply-entrenched prejudices and to take any path not already well worn.

8.31   I will conclude this chapter with an experiment much less decisive than those I have already described, but which, however, seems to me to have made more of an impression than any other on many people. If one burns a livre or 16 ounces of spirit of wine, i.e., alcohol, in a device designed to collect all the water released during the combustion, one obtains 17 or 18 ounces.* Now no material can increase in weight during an experiment, so it must be that another substance affixes itself to the spirit of wine during the combustion. And I have shown that this other substance is the base of the air, or oxygen. Spirit of wine thus contains one of the principles of water (hydrogen) and atmospheric air provides the other (oxygen). This is a new proof that water is a composite substance.

* See the description of this device in the third part of this work.

8.30 *phenomena such as the fermentation of spirits ... depend on the decomposition of water*: In Chapter 13, not included in these selections, Lavoisier will moderate his assertion that water decomposes in vinous fermentation (winemaking).

8.31 *a device designed to collect all the water released*: Lavoisier's lengthy description is not included in this selection; but he refers to the apparatus depicted here, in which alcohol (or other combustible liquid) in the reservoir L, at the lower left, supplies a lamp at the base of the vertical chimney. When the liquid burns, its gaseous products are routed through the spiral "worm." If those gases include water (as they invariably do), that water condenses and collects in the flask P at the lower right.

Fig. 5.

Excerpt from Chapter Seventeen

17.18   ... [The neutral salts constitute] a vast field, open to the zeal and activity of young chemists. But as I complete this chapter, I would like to recommend to those who possess the courage to undertake this work to strive to do well rather than a great deal; to ascertain, at the outset, the composition of acids by many precise experiments, before focusing on the composition of neutral salts. An edifice must be built upon solid foundations if it is to brave the ravages of time. And the march of chemistry, at this stage, would be impeded if it were to proceed from experiments which are neither sufficiently precise nor sufficiently rigorous.

17.18 *strive to do well rather than a great deal*: In addressing this advice to "young" chemists, does Lavoisier mean to imply that, although his own generation (Lavoisier was 44 when he wrote these words) has been granted a long and sweeping vision, from now on only patient, careful, and meticulous effort will make that vision a reality? If, indeed, *doing well* truly conflicts with *doing much*, Lavoisier's counsel has much to recommend it. But perhaps Lavoisier's own achievements testify that the very attentiveness and conscientiousness which characterize the *Treatise* may further one's accomplishment more than they restrain it.

Appendix:

Pre-Revolutionary French Weights and Measures

Units of Length

point	$1/12^3$ pied	.188 mm or .074 in	
ligne	$1/12^2$ pied	2.256 mm or .089 in	corresponds to English *line*
pouce	$1/12$ pied	27.07 mm or 1.066 in	corresponds to English *inch*
pied du roi *or* pied		32.48 cm or 1.066 ft	corresponds to English foot; known in English as the *Paris foot*
toise	6 *pied*	6.394 ft	corresponds to English *fathom*

Units of (Liquid) Volume

demiard	1/4 *pinte*	238 ml or 1/2 US pint
chopine	1/2 *pinte*	476.1 ml or 1 US pint
pinte		952.1 ml or 2.01 US pint
quade	2 *pinte*	1.904 L or 1/2 US gal
quartaut	72 *pinte*	68.55 L or 36 US gal
pouce cube	1/48 *pinte*	19.84 ml or .671 US fl oz
pied cube	36 *pinte*	34 .28 L or 9.056 US gal

Units of Mass

prime	$1/24^3$ once	2.213 mg
grain	$1/24^2$ once	53.11 mg
gros	1/8 *once*	3.824 g or 0.135 US oz
once	1/16 *livre*	30.59 g or 1.079 US oz
livre		489.5 g or 1.079 US lb
quintal	100 *livre*	48.95 kg or 107.9 US lb

Degrees of Temperature

The Réaumur temperature scale designates the freezing and boiling points of water as 0° and 80°, respectively. The chart displays it in comparison to the Celsius and Fahrenheit scales.

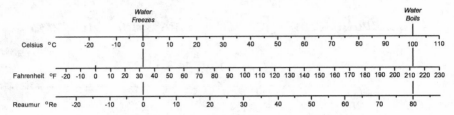

Glossary of Names and Terms

Academy: Refers to the French Academy of Sciences (*Académie des sciences*), a learned society founded in 1666 by Louis XIV. One of the earliest of such academies, it was at the forefront of scientific developments in Europe in the 17th and 18th centuries.

acid: "Having the taste of vinegar," from Latin *acidus*, "sour, sharp, tart." Lavoisier notes the parallel with Greek ὀξύς, "sharp."

aeriform: Similar to atmospheric air in having no fixed volume (unlike a liquid) and no fixed shape (unlike a solid).

affinity: Refers to the tendency of a substance to combine by chemical reaction with another substance. *See also* Bergman.

alembroth, salt of: A substance supposed by some alchemists to be a universal solvent, later used as an antiseptic. Also called "salt of wisdom," "philosopher's salt."

algaroth: A white solid, formerly used in powdered form as an emetic.

amalgam: "Combination of mercury with another metal," from Latin *amalgama*, "alloy of mercury, especially with gold or silver"; perhaps an alteration of Latin *malagma* "poultice, plaster," or Greek *malagma*, "softening substance," from *malassein* "to soften." More generally, any blend or conglomeration.

aurora borealis: Northern Lights (lit. "northern dawn"). Auroras seen from sufficiently northern locations may be directly overhead, but viewed from lower latitudes they illuminate the poleward horizon as a greenish glow, or sometimes a faint red, as if the Sun were rising from that direction. The analogous phenomenon in the southern hemisphere is *aurora australis*.

Bergman, Torbern Olof (20 March 1735 – 8 July 1784): A Swedish chemist and mineralogist noted for his 1775 *Dissertation on Elective Attractions*, containing the largest chemical affinity tables ever published. *See also* affinity.

Berthollet, Claude Louis (9 December 1748 – 6 November 1822): A French chemist who in 1785 was first to determine the elemental composition of the gas ammonia.

Boerhaave, Herman (31 December 1668 – 23 September 1738): A Dutch botanist, chemist, and physician. He was the first physician to utilize the thermometer in clinical practice.

Boyle, Robert (25 January 1627 – 31 December 1691): A natural philosopher, chemist, physicist, and inventor. With the help of his assistant Robert Hooke he constructed an improved air pump, with which he carried out experiments on vacuum and established the relation between pressure and volume of air ("Boyle's Law").

calcination: The process of thoroughly burning or roasting a mineral or metal, so as to consume or drive off all its volatile parts. Lavoisier describes calcination in Chapter Seven, second paragraph. *See also* calx.

calx: A powder or friable substance produced by calcination of a mineral or metal. Sometimes denotes the calx *lime* in particular. Plural *calces*.

carburet: A combination, but not necessarily in a fixed proportion, of charcoal with another substance, usually a metal. Steel is a carburet of iron; but its charcoal content can vary from a minute amount to a little over 2%, with corresponding alterations in its properties.

Chaptal, Jean-Antoine (5 June 1756 – 30 July 1832): A distinguished French chemist, physician, agronomist, industrialist, statesman, educator and philanthropist. Chaptal's first scientific treatise, the *Éléments de chimie* (3 vols, Montpellier, 1790) introduced the term "nitrogen" in accordance with Lavoisier's system of chemical nomenclature. Nevertheless, Lavoisier's original term *azote* remains current in French.

charcoal: Lavoisier uses the term *charbon*, translated here as "charcoal," to refer both to a simple combustible substance and a constituent of carbonic acid gas (Chapter Five) and also as we would use "charcoal" today, to refer to the residue of vegetable matter when decomposed at a high temperature in the absence of air.

colcothar: The brownish red oxide of iron which remains after the distillation of sulphuric acid from iron sulphate. Also called Crocus Martis, its finely powdered form constitutes jewelers' rouge.

compound: We translate Lavoisier's *combinaison* as "compound" when it refers to the product of a chemical reaction and as "combination" when it refers to the process of chemical reaction.

Condillac, Étienne Bonnot (Abbé) de (30 September 1714 – 3 August 1780): French philosopher and friend of Rousseau, strongly influenced by Locke. Condillac's radical empiricism was popular in the second half of the 18th century in France. He is known, among other achievements, for his contributions to the philosophy of language and for a thought experiment, in defense of his empiricist epistemology, in which he demonstrated how a statue animated by a human soul could acquire all the ideas human beings possess through sensation.

crucifers: The family of flowering plants also known as the mustards or the cabbage family. The name, meaning "cross-bearing," refers to the four petals of mustard flowers, which resemble a cross. The family contains the cruciferous vegetables, including broccoli, cabbage, cauliflower, kale, collards, turnip, Chinese cabbage, radish and horseradish.

decompose: To separate a substance into its constituent parts. At the opening of Chapter Seven, Lavoisier makes clear that since *caloric* is a constituent of gases, even elementary gases like oxygen may be said to be "decomposed."

de Fourcroy, Antoine François (15 June 1755 – 16 December 1809): A French chemist and physician who collaborated with Lavoisier, Guyton de Morveau, and Claude Berthollet on the *Méthode de nomenclature chimique*, a work that helped standardize chemical nomenclature.

de Luc, Jean-André (8 February 1727 – 7 November 1817): A Swiss geologist and meteorologist. De Luc studied the phenomena of boiling and freezing and ascertained that water exhibits its greatest density not at its freezing point but at a temperature several degrees higher.

de Morveau, Louis-Bernard Guyton (4 January 1737 – 2 January 1816): A French chemist and politician who, along with Lavoisier, Berthollet, and de Fourcroy, published the *Méthode de nomenclature chimique* in 1787.

distillation: A process of separating the component substances from a liquid mixture by selective evaporation and condensation, made possible by differences in the volatility of the mixture's components. Distillation is properly a physical rather than a chemical process; but chemical reactions are sometimes unavoidable, as Lavoisier notes in Part III (not included in these selections).

docimasy: Close examination of a person or substance in order to determine nature, quality and characteristics. In ancient Greece, the process of evaluating aspirants for public office or citizenship. In metallurgy, the process of assaying metallic ores.

dry way: *see* humid way.

effervescence: "the action of boiling up," from Latin *effervescentem*, present participle of *effervescere*, "to boil up, boil over," from *ex-* "out" + *fervescere*, "begin to boil," from *fervere* "seethe, boil" (hence *brew*). *See also* ferment.

effloresce: "To flower out." In chemistry, the term denotes migration of a salt to the surface of a porous material, where it forms a coating.

element: Lavoisier uses the terms "element," "principle," and (less frequently) "simple substance" more or less interchangeably. The term "element" is usually preferred when Lavoisier wishes to emphasize that the substance is, so far as he can tell by means of analysis, irreducible; whereas the term "principle" is usually preferred when he wishes to emphasize the fact that the substance constitutes the essence of a particular type of compound.

ferment: Intransitive verb, from Latin *fermentare*, "to leaven, cause to rise," from *fermentum*, "substance causing fermentation, leaven, drink made of fermented barley"; perhaps derived from *fervere*, "to boil, seethe"; hence *brew*. Lavoisier describes the process of fermentation at the beginning of Chapter Thirteen (not included in these selections). *See also* effervescence *and* yeast.

flask: From Latin *flasco*, "container, bottle."

French thermometer: *See* Réaumur.

gallnuts: The hardened secretions which some plants (notably oak and sumac) produce when irritants are released by certain species of insect larvae. Oak gallnuts have a brown, globular, nutlike appearance.

gas: Matter in the state of an elastic aeriform fluid. In Chapter Four, Lavoisier characterizes gases as being saturated in the highest degree with caloric. The name, originated (in Dutch) by van Helmont, derives from Latin *chaos*.

Geoffroy, Étienne François (13 February 1672 – 6 January 1731): A professor of medicine and chemistry and a member of both the Royal Society and the Academy of Sciences, he played a major role in the transmission of scientific discoveries between England and France. He is known for the table of chemical affinities he published in 1718.

glasswort: *see* kali.

graphite: A crystalline form of carbon, formerly called *black lead* or *plumbago*; both these names arose from confusion with the similar-appearing lead ores, particularly galena. Abraham Gottlob Werner coined the name *graphite* ("writing stone") in 1789 in an attempt to clear up the confusion between molybdena, plumbago and black lead after Scheele proved in 1778 that they are different minerals.

humid way: In alchemical literature, the terms *humid way* and *dry way* refer primarily to methods for creating gold. For Lavoisier, however, they apply, respectively, to chemical operations that involve, or do not involve, water.

Ingenhousz, Jan (December 8, 1730 – September 7, 1799): A Dutch physiologist, biologist and chemist, best known for having discovered photosynthesis and the essential role played in it by light. He also discovered that plants, like animals, carry out respiration at the cellular level.

kali: A common seaside plant, also known as glasswort and saltwort; it has thick, fleshy, cylindrical leaves, each bearing a stout thorn. Its ashes yield an important ingredient used in making glass and soap.

Kunkel's phosphorus: Elemental "white" phosphorus. First produced by Hennig Brandt in 1669, it emits a faint glow when exposed to the oxygen in air and was accordingly named φωσφόρος, "light-bearer." Johann Kunkel (or Kunckel) afterwards produced it in commercial quantities, so that the material became widely known as "Kunkel's phosphorus." Phosphorus is highly flammable—a property utilized by Lavoisier in the experiments of Chapter Five.

Laplace, Pierre-Simon (23 March 1749 – 5 March 1827): A vastly influential French scientist whose work encompassed mathematics, statistics, physics and astronomy. From 1782 until 1784 Laplace and Lavoisier carried out joint investigations on the specific heats and the heats of combustion of various substances.

lime: An alkaline earth, the chief ingredient in masonry mortar. It is obtained by exposing limestone to a red heat; carbonic acid is driven off, and the brittle white solid that remains is pure lime, also called *quicklime*. Lime is strongly caustic and readily combines with water to form *slaked lime*, releasing considerable heat in the process.

limestone: A rock consisting chiefly of carbonate of lime; it yields lime when fired to red heat. *Marble* is limestone in crystalline form.

lixiviate: To extract a soluble constituent by washing and filtering. From Latin *lixivius*, "made into lye," from *lix*, "ashes, lye."

lute: A type of clay or cement used to stop holes, seal joints, etc. From Latin *lutum*, "mud, dirt, mire, clay."

Macquer, Pierre-Joseph (9 October 1718 – 15 February 1784): An influential French chemist opposed to Lavoisier's views. He published the *Dictionnaire de chymie* in 1766 and was also active in French industry, particularly the development of porcelain.

magazine: A place for storing goods, especially military ammunition; from Arabic *khazana*, "to store up." (The English sense of "periodical journal" dates from 1731.)

marble: *See* limestone.

metal: From Latin *metallum*, "mine, quarry, mineral, what is got by mining," from Greek *metallon*, "metal, ore," probably related to *metallan*, "to seek after." Compare Greek *metalleutes*, "a miner," *metalleia* "searching for metals, mining."

Methodical Encyclopedia (*L'Encyclopédie méthodique ou par ordre de matières*): An expanded and updated version of the better known *L'Encyclopédie, ou dictionnaire raisonné des sciences, des arts et des métiers* edited by Diderot and d'Alembert. Many of the chemists to whom Lavoisier refers in the *Elementary Treatise* contributed to this Encyclopedia, which ran to over 200 volumes published over the course of fifty years, beginning in 1782.

Meusnier de la Place, Jean Baptiste Marie Charles (19 June 1754 – 13 June 1793): A French mathematician, engineer and Revolutionary general. He is best known for Meusnier's theorem on the curvature of surfaces. He worked with Lavoisier on the decomposition of water and evolution of hydrogen.

Monge, Gaspard (9 May 1746 – 28 July 1818): A French mathematician, best known for introducing what is now called orthographic projection of 3-dimensional curves. But Monge was also active in metallurgical and chemical researches that included the composition of nitrous acid, iron and steel.

noxious gas: Lavoisier uses *mofète* and *mofette*, which we translate as "noxious gas," in a manner consistent with Macquer's definition in the *Dictionnaire de chymie* to mean "unwholesome exhalations or vapors"; though he sometimes uses the term to refer, in particular, to the non-respirable part of the air—or, in his terminology, azote gas or, alternatively, azotic gas.

oil: In Chapter Ten (not included in these selections) Lavoisier describes oils as resulting from "hydrogen and charcoal uniting without the caloric bringing hydrogen into a gaseous state."

oxide: The first degree of oxygenation that does not form an acid. Lavoisier discusses this usage in Chapter Seven.

Papin's machine: Refers to Papin's "digester," a kind of pressure cooker designed to extract fat from bones, rendering them brittle and easily ground into bone meal.

phagedenic water: From Greek *phagedainikos*, "flesh-eating." A preparation from quicklime and corrosive sublimate, formerly used to remove fleshy growths.

plumbago: "Black lead," alternatively, any of several types of lead ore; from Latin *plumbum*, "lead." *But see* graphite. In paragraph (0.29) above, Lavoisier lists "plumbago" as one of several combinations of charcoal and iron; but he is practically alone in that usage.

pneumato-chemical apparatus: A device that enables the separation of different types of gas in order to weigh them. Lavoisier describes this equipment in Part III, not included in these selections.

pompholix: Zinc oxide, especially in the crude form produced by sublimation in a furnace.

potash: A water-soluble substance extracted from the ash produced when vegetable matter is burned in air. Lavoisier discusses it in Chapter Sixteen, not included in these selections.

quicklime: *See* lime.

radical: Adjective from Latin *radicalis*, "of or having roots," from Latin *radix* (genitive *radicis*), "root" (hence *radish*). As a noun, the root part of a word (linguistics); the root principle of a compound substance (chemistry).

Réaumur, René Antoine (28 February 1683 – 17 October 1757): A French scientist who contributed to many different fields, especially the study of insects. In 1730 he introduced a scale of temperature based on the expansion rate of alcohol; but long before Lavoisier's time Réaumur's original scale was redefined according to the freezing and boiling points of water; it is styled "the French thermometer" by Lavoisier in Chapter One.

red heat: In Chapter Twelve (not included in these selections) Lavoisier writes: "I will sometimes make use of the expression 'red heat.' Although it does not express a determinate degree of heat, it does, however, express a degree of heat much greater than that of boiling water."

resin: From Old French *resine*, "gum, resin," and Latin *resina*, from Greek *rhetine*, "pine gum." Hence also *rosin*.

retort: A vessel with a bent neck; from Latin *retorta*, literally "a thing bent or twisted," from past participle stem of Latin *retorquere*.

salt: At the opening of Chapter Sixteen (not included in these selections) Lavoisier restricts the term "salt" to "composites formed by joining a simple oxygenated substance to a given base."

saltwort: *see* kali.

Saussure, Nicolas-Théodore de (14 October 1767 – 18 April 1845): A Swiss chemist and student of plant physiology who made discoveries in plant-derived chemicals and in the study of photosynthesis.

spirit of wine: Lavoisier mentions this term, in addition to "eau-de-vie" and "aqua vitae," as referring to pure alcohol; although these terms were, and still are, often used to refer particularly to brandy (distilled wine), a liquor that is not usually pure.

Stahl, Georg Ernst (22 October 1659 – 24 May 1734): A German chemist, physician and philosopher. He was a supporter of vitalism, and until the late 18th century his writings on phlogiston provided the accepted explanation of combustion.

sulphuric ether: A volatile and flammable liquid, made by distilling ethanol with sulphuric acid, and widely used as a solvent. It is the "ether" formerly (from about the middle of the 19th century) used in surgery as a general anaesthetic; but its anesthetic properties were known to Paracelsus in the 16th century.

turpeth mineral: A lemon-yellow alkaline sulphate of mercury, obtained from common sulphate by washing with hot water. Also prepared from the leaves of the turpeth plant, it was formerly used as a cathartic.

van Helmont, Jan Baptist (12 January 1580 – 30 December 1644): A Flemish chemist, physiologist, and physician. He coined the word "gas" (from Greek *chaos*) and was the first to recognize that there are gases distinct in kind from atmospheric air.

Vandermonde, Alexandre-Théophile (28 February 1735 – 1 January 1796): A musician, mathematician and chemist. He performed experiments with Lavoisier on the great frost of 1776, and published two papers on the manufacture of steel with Monge and Berthollet.

vitriol: Any of a number metallic compounds having a glassy (Latin *vitrum*) appearance. In Chapter Five, Lavoisier recounts the connection between vitriol of iron ("green vitriol") and sulphuric acid; similar experiments would indicate that the common element in "vitriols" is sulphur.

volatile: "Fine or light," also "evaporating rapidly," from Latin *volatilis*, "fleeting, transitory; swift, rapid; flying, winged," from past participle stem of *volare*, "to fly." In Middle English, *volatiles* denoted birds, butterflies, and other winged creatures.

water of crystallization: Refers to the water that is necessary for the formation of some crystals, and which is incorporated in the crystal structure.

yeast: The name comes from Old English *gist* or *gyst*, and from the Indo-European root *yes-*, meaning "boil", "foam", or "bubble." In Chapter Thirteen (not included in these selections), Lavoisier describes yeast as initiating fermentation by "disturbing the equilibrium" of the components of sugar. Not until 1857 was it proved (by Louis Pasteur) that yeast's role in fermentation was that of a living organism. *See also* ferment.

Index

proceeds from the known to the
unknown 2
structure of 1

sensation
of cold and heat 27
requires movement 27

T

temperature
absolute cold 13
of boiling water 20
of human body 21

themometer
principle of 27

V

van Helmont, Jan Baptist 52

vapor. *See* gas

vital air. *See also under* atmospheric air
weight of 48

vitriol 68

vitriolic acid 69. *See also* acid: of sulphur:
sulphuric

W

water
aeriform state of 51
composed of hydrogen and oxygen 85,
90
decomposition of 82
releases hydrogen 83
releases oxygen 84
formerly thought to be an element 80
recomposition of 86, 87, 88
simple distillation of 81
solid state of 51

weight
constancy of 57, 61, 64, 65, 89